How To Diagnose and Repair **Automotive Electrical Systems**

How To Diagnose and Repair **Automotive Electrical Systems**

Tracy Martin

To Leslie, whose love and advice is a gift.

First published in 2005 by Motorbooks, an imprint of
MBI Publishing Company, 400 First Avenue North,
Suite 300, Minneapolis, MN 55401 USA

Motorbooks titles are also available at discounts in bulk
quantity for industrial or sales-promotional use. For
details write to Special Sales Manager at MBI
Publishing Company, 400 First Avenue North, Suite
300, Minneapolis, MN 55401 USA.

To find out more about our books, join us online at
www.motorbooks.com.

On the front cover:
Main: Sometimes a lab scope is the only way to
determine if an electrical component is doing its job.
The waveform for one of the eight fuel injectors on
this Ford F150 looks good. **Small:** These components
may appear different, but they're all just load devices
with resistance to electron flow.

On the back cover: A relay has been added to a
driving light circuit. The relay now controls the high
amperage load that the driving lights need to operate.
For more on this, see Chapter 8.

ISBN 978-0-7603-2099-0

Editors: Jennifer Johnson and Peter Bodensteiner
Designer: Chris Fayers

Printed in China

CONTENTS

SECTION III: ELECTRICAL SYSTEMS

PREFACE

While the automotive electrical field has always fascinated me, it has taken me about 25 years to get to the position of writing this book. I have spent years working in my garage on all makes of vehicles. It has oftentimes been a long and arduous process of trial and error, discovering what works and what doesn't, how to fashion "short-cuts," and most importantly, how to break it all down into simple, easy-to-understand "sound bites" of information that a reader or beginning mechanic can quickly understand and digest. Fortunately, what has made this process somewhat easier is that I have been taught by some of the best people in the fields of automotive repair and training. Many of the instructional methods used to convey technical information in this book have been borrowed from colleagues, friends, and fellow mechanics and then elaborated upon. However, throughout my experiences I have often been struck by the notion that the related areas of automotive electricity, electronics, and diagnostic testing are needlessly complicated and presented in a confused manner. I have endeavored to correct this imbalance in this book.

In the early 1980s I worked as a lab technician at a large turbocharger manufacturer in Southern California. This company was one of the few pioneering innovators in design and construction of turbochargers for the diesel and automotive markets. It also had the largest engine dynamometer facility on the West Coast. Many auto manufacturers introduced the advantages of turbocharging an engine in the early 1980s, and as a technician in the right place at the right time, I was fortunate enough to see new and developing automotive technology years prior to its incorporation into production vehicles. This environment was the setting in which I received an introduction to some of the earliest electronic engine management systems, their modes of operation, and their inherent "teething" problems.

Along with fellow technicians, I had the opportunity to install test engines into various dyno test cells (soundproof rooms), mounting all the electronics on a plywood board next to the engine. As engine testing progressed, the computer-controlled carburetor or fuel injection system would inevitably "crash and burn" and have to be repaired. Since there were no service manuals available (they aren't written preproduction), automotive engineers were the

only source of information for how the systems were supposed to work. And more often than not, the response from engineers to queries for more information about why something didn't work out as expected went something like this: "The system couldn't possibly have malfunctioned; it's made using state-of-the-art electronics, designed by a qualified electrical engineer. You must have installed it incorrectly." Although this was not the most ideal environment in which to learn automotive electronics, some technical knowledge rubbed off on me with the help of fellow technicians.

Some years later, the widespread introduction into the automotive industry of carburetors with wires coming out of them and electronic fuel injection systems gave me the opportunity to teach fellow mechanics (now called technicians) what I had learned about how to diagnosis and repair these systems. I taught classes for many nationwide corporations, including Sun, Allen, Nissan, and Snap-On.

Typically, these instructions occurred in the evenings, after everyone had already put in a full day working on cars. Too often, I was faced with a group of ill-fated students all wishing they were home eating dinner instead of sitting in class. Keeping these students awake, interested, and open to learning was a challenge, to say the least. Consequently, over the years I gained invaluable experience and learned, out of necessity, innovative ways of imparting information and keeping things moving during class. As a result, I have been able to incorporate many of those techniques and ideas into this book.

In addition to teaching, I had the great fortune to spend quite a few years working and consulting at a unique and interesting business called Automotive Data Systems (ADS) in California. ADS is a cutting-edge company that provides a telephone automotive diagnostic "hotline" that technicians can call in to receive immediate, real-person, real-world advice on how to diagnose and repair a specific problem on a car as it's being worked on. The automotive database compiled, referenced, and stored at this facility is truly amazing, having (at last count) over 30,000 records, or "tech-notes," pertaining to specific years, makes, and models of vehicles and their drivability-related problems. As a result, ADS technicians have been consistently able to give correct advice to customers over 95 percent of the time. With this much confidence in that database and the daily hands-on training I received in diagnostics and troubleshooting, I was able to synthesize a great deal of information about electrical automotive systems and their diagnosis and repair. I am forever indebted to the great bunch of guys I met at this company who have always generously shared (and continue to) their knowledge and experience with me. Unfortunately, the company, although still in business, has been swallowed up by a large automotive aftermarket "solutions" provider; many of the employees now work for a large Korean car manufacturer somewhere in Southern California. Hopefully, they've all found diagnostic nirvana there.

Finally, for the last 25 years of my life there has been my wife. On our first date, she showed up at my door only to find me working in my garage on my recently acquired 1964 Corvette. It should have been a portent for her of things to come. However, she was not dissuaded. For this, I am forever indebted to her, especially for her continuing efforts to pump some life into me and make me funny and accessible to humans. I don't know where I'd be without her—probably somewhere else sucking engine juice for life support and wishing I had a wife just like her.

If all these experiences have taught me anything, it's that the more I think I know, the more I realize I don't know much at all. So I'd like to thank some of the people who have helped me over the years, especially with the writing of this book.

Chief and foremost, I'd like to thank "Teck." He was my master teacher during my short tenure as a high school auto-shop teacher and also the founder of ADS. His technical and personal advice over the years has been an invaluable source of inspiration, development, and practical growth, and without him I am sure I would be something less of the person I am today. I'd also like to thank both Curt Moore and Dave Bellaver, who allowed me access, through their contacts, to much of the technical information contained in this book. I'd also like to thank especially Mike McElfresh, a former coworker, technical scribe, and overall automotive scholar at ADS. Without his technical editing, and invaluable input, some of the stupid things I originally wrote would have wound up printed in these pages for you to laugh at. Fortunately, his generous gift of time and his always-willing-to-help attitude has (hopefully!) saved me from professional embarrassment. Thanks, Mike, for all your help. Lastly and again, there is my wife, whose incredible patience, astute editing skills, and sharp-edged advice (OUCH!) made this a better book. She helped me avoid writing about voltmeters or alternators that "speak" or ground path returns that go nowhere. I'd like to thank her from the bottom of my heart and promise her that she doesn't have to read this book "just one more time."

In closing, I'd just like to say that I hope you, the reader, are able to gain some practical skills and knowledge, which will help increase your confidence when faced with automotive electrical challenges.

— *Tracy Martin*

SECTION I
THEORY

This first section is intended to provide a bare-bones explanation of general electrical theory and how basic direct current (DC) electricity operates in an automobile. It's not important to understand electricity inside and out—a subject many books cover in excruciating detail—but it is important to have a basic understanding of how to apply a practical working knowledge of electricity in order to diagnose and repair electrical malfunctions that show up in your car or truck.

An in-depth examination of electrical theory is far more complex and cumbersome than the practical "hands-on" premise offered in this book. Therefore, apologies are offered in advance to readers with more than a working knowledge of electricity or electronics since this book takes certain liberties and shortcuts with electrical science. For all electronically challenged mortals, Chapter 1 on Ohm's Law is a hot-rod version of how 12-volt automotive electrical systems operate. Then, Chapter 2 on voltage drop testing cuts to the chase by showing how to apply the information learned in the context of solving real automotive problems. In fact, all subjects covered in this book relate in some way to basic theories discussed in chapters 1 and 2. So, if brain-fade starts to set in while reading more detailed sections on wiring diagrams or electronic fuel injection diagnosis, a revisit to these chapters for an electrical theory "tune-up" may prove helpful.

CHAPTER 1
OHM'S LAW

Because billions of electrons flowing through a wire at the speed of light are difficult to see and—for most people—hard to even imagine, electronics and electrical repair are areas of vehicle maintenance most people shy away from. Unlike disassembling and cleaning a carburetor, changing a flat tire, or bolting accessories onto your vehicle, repairing electrical systems is a truly cerebral endeavor; however, it is not impossible or even difficult. Electrical systems may seem perplexing while watching a seasoned automotive technician with electronics savvy diagnose an electrical problem in a vehicle, especially when you think you may have to repeat the process when faced with an electrical nightmare at a later point. However, a little secret allows you to easily repeat skills needed to diagnose a starter motor that goes "click," a dim headlight, an engine that starts and dies, or to find and fix any number of other mysterious problems. The secret is practice.

Most people understand the need for practice, especially when it comes to sports; playing baseball, riding motorcycles, shooting pool, bowling—anything requiring a specific skill set goes a lot smoother with experience. For example, suppose you haven't picked up a baseball bat in a year and it's your turn at bat in a game, there's a good chance you'll strike out. As everyone knows, a little practice before a game goes a long way toward ensuring success. Likewise, the day your vehicle fails to start also should not be the first time you switch on your new digital voltmeter. Simply having basic electrical knowledge is not enough; you must practice applying that knowledge by using a multimeter on an operating circuit. This will provide a better-than-average chance of hitting the ball the first time (electrically speaking) when something goes wrong with your car. And it's easier than you think.

Applying electrical theory to the real world is simply a matter of knowing what reading to expect from a volt/amp/ohmmeter display connected to a working circuit before connecting it to a problem circuit. Anticipating what the reading should be and understanding what the numbers mean allows visualization of an otherwise invisible problem. By practicing on operating circuits with known values, you'll gain the necessary confidence to figure out what to look for in a circuit that isn't doing what it's supposed to. Let's start by dissecting a common 12-volt DC circuit—it doesn't get much simpler.

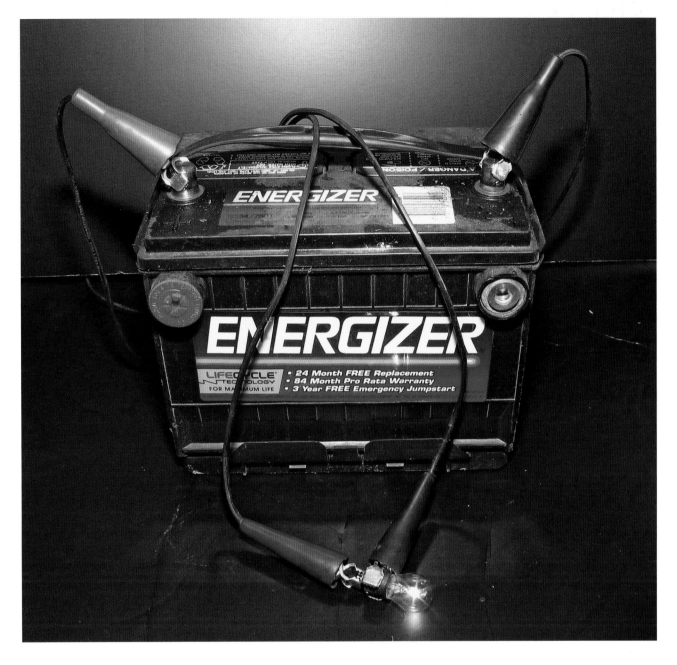

A simple 12-volt battery, wires, and light bulb make up a basic DC circuit.

"THREE THINGS" ABOUT 12-VOLT DC CIRCUITS

Everything electrical in a vehicle is part of a circuit. Circuits are simply layouts, or designs, of how electrical components are powered and controlled. Electrical components found in automobiles are usually divided into categories of circuits (though not always). For example, a lighting circuit is composed of headlights, taillights, running lights, and interior lights; a charging circuit includes an alternator (or generator), voltage regulator (if used), and battery; typical fuel injection circuits have an electronic control module (ECM) and various sensors and actuators.

Within each system are individual circuits that control specific electrical components. Headlights and taillights are part of the lighting system, but each operates via a separate circuit within that system. This system within a system, or subsystem, creates a big stumbling block for many "shade tree" mechanics and professional technicians alike who are faced with problems in automotive electrical systems. In order to diagnose an electrical problem, a nonoperational

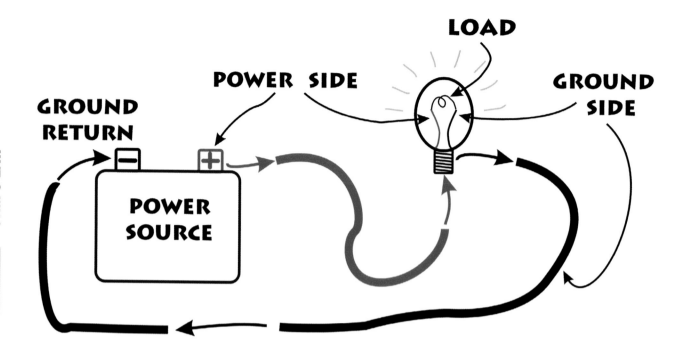

Fig 1-1. *Three elements of a 12-volt DC circuit: (1) the battery's positive terminal and red wire are the* power source, *(2) the load or* load device *is the filament inside the light bulb, and (3) the black wire is the* ground return.

circuit must be separated and isolated from the overall larger system of which it is a part, as well as from other operating circuits within that same system.

When faced with an automotive electrical repair, most people typically turn to manufacturers' wiring diagrams for help. They wrongly assume that since the diagrams provide a blueprint of the electrical system, they will thus help with the identification of specific inoperable circuits. However, this approach is like looking for a needle in a haystack. Manufacturers' wiring diagrams don't isolate or identify inoperable circuits; in fact, since they can show the entire lighting system with all its circuits, or worse, the complete electrical system for the whole vehicle, this approach can prove daunting unless you know what to look for.

The ability to identify and isolate a circuit allows you to simply connect a voltmeter properly and anticipate the respective readings. This is not hard or intimidating if you understand the "Three Things" that make up all 12-volt DC circuits. When any one of the "Three Things" goes missing in action, the circuit stops working. While this may seem obvious, it is far less so when looking at a complex wiring diagram or actual wiring harness under the hood. However, when you know what to look for, these "Three

Things" are easily identified as the primary components of every 12-volt DC electrical circuit.

The "Three Things" listed—power source, load device, and ground return—are all necessary and must be present in a circuit in order for it to operate.

Power Source

Every electrical component must have a *power source* in order to operate. All electrical energy needed for the circuit to do its job is provided by a power source. In order for electricity to move along a wire, subatomic particles called electrons (invisible to the naked eye) interact to transfer energy from one point to another; they provide power for a circuit. Starting at a battery's positive terminal, electrons are pushed through the circuit. Any problems with connections on the power side of a circuit will affect the entire circuit. This seems like a no-brainer, but oftentimes technicians and home mechanics spend countless hours trying to discover why something won't work only to find out later that a simple blown fuse is the cause.

The battery and/or alternator/generator are the chief power sources for all electrical and electronic circuits in a vehicle. In addition, wires connected to either a battery's positive terminal or an alternator's output terminal are considered a power source. Consequently, relays, fuses, junction blocks, and fusible links also provide power to electrical components because they connect to the battery's positive terminal. Additional terminology used to refer to power sources includes: plus side, power side, hot or hot side, positive (+), and battery.

Load Device

A *load device* is any component that uses up voltage or has resistance to electrons flowing through a wire. Most load devices simply amount to nothing more than electrical conductors of various lengths, sizes, and shapes. For example, motors, relays, lights, solenoids, coils, spark plugs, and computers (black boxes) are all load devices with some resistance. The hundreds of load devices appearing in wiring diagrams all perform some type of useful work and are included as part of the design of the circuit in which they function.

However, there is one type of load device we can all live without: unwanted load devices. These run the gamut from corroded or loose connections to frayed sections of wire to dirty contacts inside switches or relays. Unwanted load devices have resistance to electron flow, use up voltage unnecessarily, and have an undesirable effect on electrical circuits. Worst of all, they don't show up on wiring diagrams, so you have to find them yourself! (Chapter 2 on Voltage Drop Testing will show you how.)

Ground Return

A *ground return* provides a route for electrons (electricity) to return to the battery after use by a load device. These can be wires, metal body panels, the engine block, the transmission, or a vehicle frame. Other terminology often used

Two power sources found in all vehicles—the storage battery and alternator (or generator in older vehicles).

These components may appear different, but they're all just load devices with resistance to electron flow. Courtesy Younger Toyota

to refer to a ground return includes: ground, cold, earth, or negative (-). This sequence of electron flow—power, load device, and ground return—is known as a complete circuit. If any of the "Three Things" is disconnected, the circuit is broken and rendered incomplete.

Because electrons are invisible when flowing through a circuit, it's hard to get an idea of what's going on inside the system. Consider a basic 12-volt circuit consisting of a battery, light bulb, and wires. The only visual confirmation that the circuit is operating is that the bulb is on. If everything is connected and the light bulb is off, one of the "Three Things" is missing and the only confirmation a technician receives that something is wrong is that the bulb will not turn on.

Let's think of this concept another way. Water flowing through a hose is a user-friendly way to conceptualize what's happening inside an operating electrical circuit. Visualize a tank full of water with an internal pump. There is an inlet and outlet on the tank connected via hoses to a load device. When water is pumped out of the tank under pressure, it's sent to a load device that does some form of work. After the

energy from the pressurized water is extracted via the load device, a return hose (ground return) sends the water from the load device back to the tank where the sequence starts over again.

For this process to work continually, water must be returned to the tank at the same rate it's pumped out. The flow of electrons through an electrical circuit works in a similar manner. Starting from the circuit's power side, electrons flow to a load device where they provide energy for some type of work to be accomplished. From the load device they flow through the ground-return wire back to the battery.

"THREE MORE THINGS" ABOUT 12-VOLT DC CIRCUITS

You knew it couldn't be that simple! Only three electrical concepts to keep track of? Don't worry; there's not too much more.

The power source, load device, and ground return are physical objects that can actually be seen and touched. In

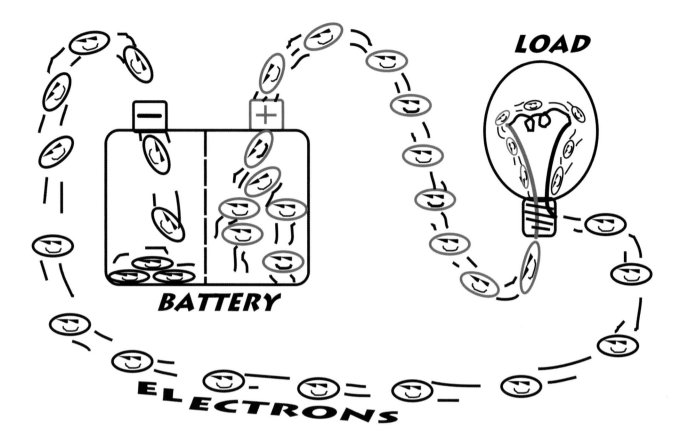

Fig 1-2. *Deliriously happy working electrons are pumped, or pushed, out of the battery and travel through a load device back to the battery via a ground return—thus, making a complete circuit.*

addition to these three physical objects, there are three basic, yet less tangible, concepts or principles for a standard 12-volt DC electrical system. Understanding their significance and interaction is just as important as understanding how the three things form a circuit. Fortunately, to help out so you're not operating totally in the dark, the net effect of the interaction of the three more things can be observed.

The "Three More Things" are: voltage, amperage, and resistance.

Think of voltage as electrical pressure, amperage as the amount of electricity used in the circuit, and resistance as restriction on the flow of electrons through the circuit. These three concepts represent the electrical values of what's actually occurring inside an operating circuit. Having a clear idea of how they interrelate provides a concrete image of what's right or wrong with a DC electrical circuit.

Voltage

Voltage can be thought of as pressure needed to push electrons from a battery's positive post, through a load device, and back to the negative terminal. Voltage, or electrical pressure, is similar to pressurized air produced by an air compressor. The compressor forces air into a tank where it is stored as an energy source. After connecting an air-powered tool to the tank and squeezing the trigger, the tank's high-pressure air is pushed from the tank through the air hose into the tool so the tool can perform work. The higher the air pressure, the more work the air tool can accomplish. The same is true of voltage. The more voltage, or electrical pressure present, the harder and more forcefully the electrons are pushed along a wire and through the load device. There are only 12.6 volts worth of push in an automobile battery. With the engine running, the alternator raises this voltage to around 14.5 volts. With only these small amounts of push present, it's important not to lose any voltage across a connector or along a wire.

However, there is another electrical system in vehicles that develops considerably more electrical pressure than the 12.6 volts found at the battery. Ignition systems require high electrical pressure because the load device for the

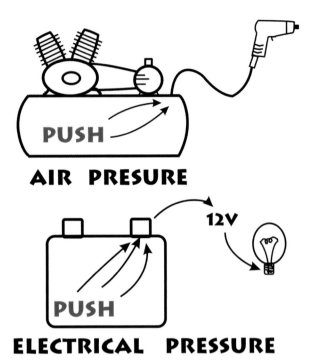

AIR PRESURE

ELECTRICAL PRESSURE

Fig 1-3. *Air pressure and voltage (electrical pressure) are similar in that more push equals more work produced by a load device.*

ignition system circuit is a set of spark plugs. A spark plug has resistance to electron flow via the small gap of air between two of its parts—the center and ground electrodes. Ignition voltage needs to be high enough to overcome the high resistance created by the air gap. The high voltage creates a spark as it jumps from the spark plug's center electrode to the ground return electrode. On older ignition systems, 25,000 volts were required to create a spark that would overcome the air gap's resistance. The output on newer ignition systems is considerably more, with many producing well over 100,000 volts. If you've ever worked on an ignition system with the engine running and been zapped by a spark-plug wire, you know what that push feels like.

Amperage

Amperage, or current, is the amount (or volume) of electricity (amps) flowing through a circuit. A starter motor with a high amperage draw has lots of electrons that must travel from the battery, through the starter, and back to the battery in a complete circuit for the starter to get enough energy to turn the engine over. As much as 250 amps are required to crank a large V-8 engine. In a starter circuit, both positive and negative battery cables have to be large

enough to allow for the unrestricted flow of electrons. In other words, the diameter of a cable in a starter system has to be large enough to offer low resistance to the magnitude of current flow traveling through the system.

Conversely, the process of illuminating a taillight requires considerably less energy. Because the bulb used in a taillight circuit is a low amperage load device, the wires from the power source to the bulb and returning back to the negative battery terminal are significantly smaller in diameter than the cables in a starter circuit. The minimal resistance found in the smaller wires will not slow the electrons substantially enough to prevent the transfer of energy necessary to light up a taillight—which is only 2 amps. The amount of amperage flow in a circuit is independent of the size of the wires used. Using a wire the size of a battery cable to construct a taillight circuit would have no effect on its operation. However, using a small taillight-sized wire as a battery cable for a starter circuit would not work. The small wire would not transfer enough electrons (high amperage) into the starter motor, and it would overheat and melt in half.

Resistance

Resistance is the restriction of electron flow in a circuit. Resistance anywhere in a circuit slows the flow of electrons. By definition, all load devices have resistance to electron flow. The relationship between voltage, amperage, and resistance was discovered about 170 years ago by Georg Simon Ohm. The theory explaining the interaction of these principles is known as Ohm's Law. The primary measurement of resistance is expressed as ohms.

Automobiles have both high and low resistive circuits. For example, a dome light bulb may have high resistance (12 ohms) to electron flow and therefore may use only a small amount of current or amperage. By contrast, a starter motor has a low resistance value (0.06 ohm) and allows a greater number of electrons (high amperage) to flow from the battery to the starter. Because resistance restricts the flow of electrons in a circuit, it affects the path (the wire) the electrons travel down. Like most of us, electricity is lazy—it takes the path of least resistance from one point to another in a circuit. For example, if one of two wires connected to two light bulbs has high resistance, the electrons will flow down the other lower resistive wire (lighting that bulb only).

RESISTANCE/AMPS RELATIONSHIP

Many electrical gremlins found in automobiles are attributable to unwanted, high resistance within a circuit. The presence of too much resistance in a circuit slows the

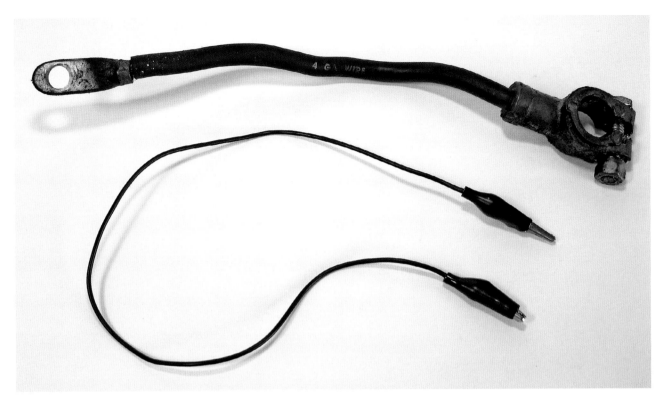

The battery cable's relatively large size allows it to carry as much as 250 amps in a starter circuit without overheating. The smaller wire will accommodate only up to 25 amps; using it in a starter circuit would melt it in half.

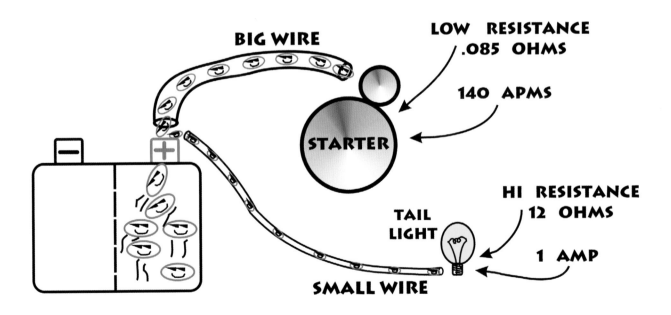

Figure 1-4. *A high-amperage, low-resistance starter circuit requires large cables to allow enough energy to reach the starter and return to the battery. However, a taillight's high-resistance, low-amperage design requires only a small wire to carry its electrical load.*

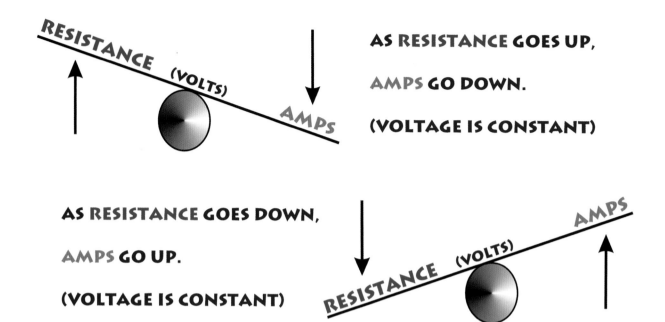

Fig 1-5. *If you remember nothing else from this chapter, remembering the relationship between resistance and amps and how they operate within a circuit is worth its weight in gold (or money!), since it will help you diagnose many common electrical problems. If amps are down, then the cause has to be unwanted, high resistance; if high amperage is present, then the circuit resistance is low.*

flow of electrons, causing low performance or nonoperation of load devices.

There is a direct relationship between circuit amperage and circuit resistance. It's critical to understand this simple cause-and-effect relationship because knowing how to apply this knowledge is key toward providing solutions to many automotive electrical problems.

Both a slow turning starter and a dim headlight are the result of an insufficient number of electrons passing through the circuit back to the battery (forming a complete circuit). Somewhere along the circuit, high resistance has blocked the flow of current. Thus, an increase in resistance causes a decrease in amperage.

However, the opposite occurs whenever resistance in a circuit is too low; a decrease in resistance causes an increase in amperage. For example, if a power source wire comes into contact with a ground return (because of a loose connection, frayed wire, or another reason), thus bypassing the intended load device, the low resistance present in the ground return allows high amperage to flow through the circuit. If the wires are too small to carry the increased amperage, they could overheat and melt, which can possibly cause a fire. Fuses are used to protect circuits when

resistance to electron flow becomes too low. The fuse heats up when amperage increases; at some point, amperage gets high enough to melt the fuse in half, causing an incomplete circuit. Fuses provide a margin of safety in circuits since a burned wire could cause an electrical fire. And a melted fuse is easier to replace than burned wires.

It's important to remember the inverse relationship between resistance and amperage. When resistance in a circuit is decreased, amperage always increases proportionally. Conversely, if a circuit has high resistance, the available amperage is decreased.

DC AND AC CIRCUITS

In a direct current (DC) circuit, the flow of amps always runs in only one direction. For ease of explanation (and in conformance with automotive publications), all diagrams used in this book depict electricity (electrons) flowing through DC circuits in one direction only—from a point of higher (positive) voltage to a lower (negative) voltage. Electron flow from positive to negative is called conventional electron theory. (In reality, electron movement at the subatomic level travels only from negative to positive in a DC circuit. See Chapter 4 for the explanation why.) However, since you can't see the

VOLTS = AMPS TIMES OHMS

OHMS = VOLTS DIVIDED BY AMPS

AMPS = VOLTS DIVIDED BY OHMS

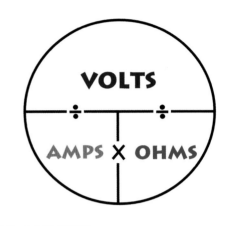

WATTS = AMPS TIMES VOLTS

AMPS = WATTS DIVIDED BY VOLTS

Fig 1-6. *These are the basic formulas needed to figure out the "numbers" in any 12-volt DC circuit. To solve for any one unknown value, you only need to know two of the three values—volts, amps, or ohms. The unknown variable can be determined by the simple mathematical equations above.*

electrons while working on a circuit, the direction in which they travel is irrelevant. The important thing to remember for practical purposes is that in DC circuits electricity always flows in only one direction. For purposes of illustration, the "hands-on" subject matter of this book uses the conventional view, based on the simple assumption (even if not correct) that electrons in a DC circuit always flow from positive to negative.

By contrast, alternate current (AC) circuits (the type typically used in American homes) reverse the direction of voltage 60 times per second. These alternating cycles of forward and backward electron flow are called hertz (Hz). This voltage reversal, which also reverses current flow, gives this type of electrical power its name—alternating current. AC circuits operate on higher voltages than DC circuits—either 120, 240, or 440 volts for an AC circuit versus only 12 or 24 volts for a DC circuit.

Be aware that some electrical components in automobiles produce only AC voltage, like the alternator and some computer sensors. However, when an alternator produces AC voltage, diodes (which are electrical one-way valves that allow current to pass in only one direction) within the alternator convert it to DC voltage before it reaches the battery. No electrical systems used in vehicles operate on AC voltage.

Furthermore, wiring configurations and components for AC current are generally not compatible with DC circuits.

OH NO, MATH?

Books on electronics are always packed full of mathematical formulas. If you get that glazed, far-away look when faced with cryptic equations like $E = I/R$, don't worry, this book keeps it simple.

When diagnosing an electrical problem on a car or truck, understanding the dynamic relationship between resistance, amperage, and voltage is critically more important than any math skills you may (or may not!) have. However, some basic calculations will come in handy if you intend to add electrical accessories to your car. For example, do you have any idea how big the wires need to be to power up that 2,000-watt stereo you bought? Should you install a larger alternator to accommodate the stereo's power so you get the sound you want? Do you know the size of the fuse needed to protect the circuit from meltdown?

The formulas listed in Figure 1-6 are designed to provide the basic math necessary to find voltage, resistance, amperage, and watts (power) in any circuit. In general, these should cover most common circuitry design needs.

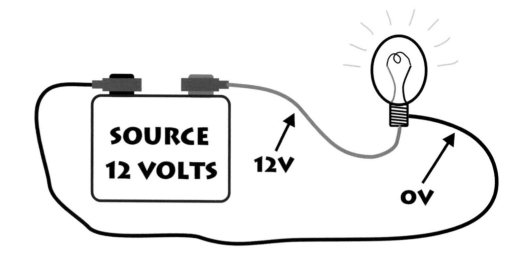

Fig 1-7. *Twelve volts are present on the red wire between the battery and light bulb (load device). The light bulb uses up all available voltage. Thus, the ground return wire reads 0 volts all the way back to the negative battery terminal.*

To determine the amount of any unknown voltage, amperage, or resistance present within a particular circuit, the values of two of the three potentials must be known—the third unknown value can be determined based on the values of the other two. Performing a few simple mathematical computations allows simple calculation of electrical loads.

To find the voltage used in a circuit, just multiply existing amps by ohms. Similarly, ohms can be calculated by dividing amps into voltage. Likewise, amps can be determined by dividing ohms into voltage. If an accessory you intend to install is rated in watts and you want to find out how many amps will be used in the circuit, just divide watts by volts. Conversely, if you want to find watts, multiply amps by volts.

RULES OF OPERATION FOR THREE TYPES OF CIRCUITS

Three types of electrical circuits are used in cars and trucks: series, parallel, and series/parallel. Series and parallel circuits are by far the most common and can be found on both new and older vehicles. Each type of circuit follows a set of operational rules that govern how the circuit works. Some rules apply to more than one type of circuit; others apply to only one type. Knowing these rules and understanding how each circuit operates gives you the advantage you need to diagnose most electrical problems.

Series Circuits

Naturally, there are "three rules" regarding how a series circuit operates.

The first rule of series circuits is: All available voltage in a series circuit will be used up by the load device.

It's important to remember this rule. By keeping it in mind, you'll be able to determine what the voltage should be at any point in a series circuit. If a voltmeter gives an unexpected result, you'll know where to look for the problem. Here's why:

Figure 1-7 shows a circuit's power source as the positive battery terminal as well as the wire connecting it to the light bulb, which is the load device. A full 12 volts from the power source are present at every point along the wire between the battery and load device (measured using a voltmeter). When 12 volts reach the light bulb, it uses up all the voltage in the circuit. Since the greedy light bulb uses all available voltage by turning on, none is present on the ground return wire or at the negative battery terminal. (Not quite true! The ground return wire also has a small amount of resistance to current flow, which causes some voltage to be present). Consequently, the ground wire will measure close to 0 volts. Chapter 2 on Voltage Drop Testing provides in-depth explanations of what to expect from voltage readings on ground return wires. For now, consider the ground wire essentially at 0 volts.

The second rule of series circuits is: When more than one load device is present in a series circuit, the individual resistance of each load device divides the available voltage, thus adding to the total resistance of the entire circuit.

Think of load devices in a series circuit as a strand of Christmas tree lights along a wire connected in series. Because each bulb has resistance to current flow, each adds to the total resistance of the entire circuit. This cumulative increase in the overall circuit resistance correspondingly decreases the amperage available for all the bulbs, thus keeping the strand of Christmas tree lights at a nice fire-safe, low-level wattage. In addition, each of the bulbs must share the available voltage because all need some voltage to light up.

Voltage in a series circuit is not a constant. It is divided between all the load devices in the circuit, based upon the individual resistance of each specific load device. It's important to remember that each load device in a series circuit requires both a power source and a ground return to operate. Because the load devices are linked together, the power source for one load device simultaneously acts as the ground return for another.

For example, consider three light bulbs connected in a series circuit as in Figure 1-8. The first bulb (bulb 1) is powered by the originating power source—a 12-volt battery. The ground return for bulb 1 becomes the power source for the second bulb (bulb 2). Bulb 1 uses 4 volts of power from the 12-volt power source (to light up) and then passes the remaining 8 volts to its ground return. Similarly, bulb 2 uses up 4 volts of available voltage from the ground return of bulb 1 (bulb 2 uses the same voltage as bulb 1 to operate). The remaining 4 volts of power pass into its ground return. Bulb 3 takes the last 4 volts and uses up all the voltage left in the entire circuit; consequently, its ground return has 0 volts.

Figure 1-8 illustrates the division of voltage along a series circuit. It also proves the first rule of series circuits still holds true; the load device uses up all available voltage since all the individual light bulbs combined cumulatively use up all the available voltage in the circuit (equivalent to a single combined load device). Thus, the ground return wire at the last load device will measure close to 0 volts. Figure 1-8 also introduces the ground symbol—three horizontal lines at the end of the ground return wire. Anywhere this symbol is placed indicates the wire is returning to the negative battery terminal and/or when a ground strap, body, or vehicle frame is used as a ground.

The third rule of series circuits is: amperage is the same at all points throughout a series circuit.

This rule is true for both negative and positive sides of a circuit. Figure 1-9 (on page 22) shows three ammeters measuring current in a series circuit and how the amperage remains constant on both the power and ground return sides of the circuit. Rule three illustrates the simple concept (but unwanted result!) that a bad wire or poor connection (unwanted resistance) will affect circuit amperage no matter which side of the circuit it's located on.

Parallel Circuits

Nearly all electrical circuits designed for cars or trucks are parallel circuits. Fortunately, the rules for parallel circuits

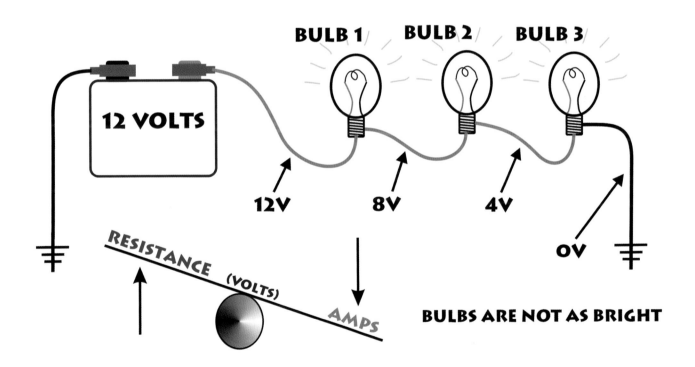

Fig 1-8. *Each 12-volt bulb (load device) has a given, identical, individual resistance, resulting in equal sharing of the overall available voltage. The difference in voltage between the power source and ground return for each bulb is 4 volts. Consequently, each bulb has only 4 volts of electrical pressure with which to operate. Since all the bulbs are designed to operate on 12 volts and not 4 volts, this shortfall in voltage for each bulb causes all of them to be dim.*

OHM'S LAW

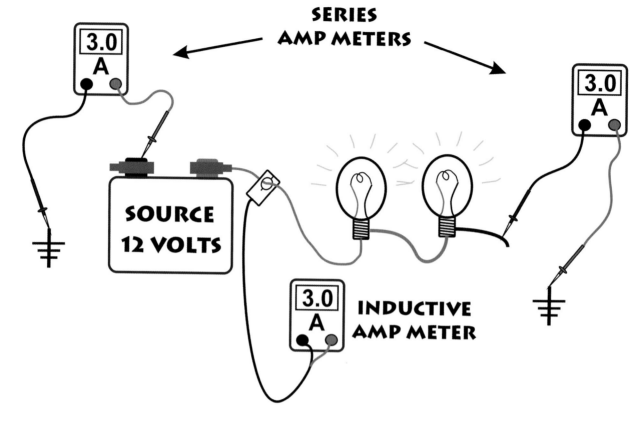

SERIES AMP METERS

INDUCTIVE AMP METER

Fig 1-9. *Amperage will be the same on both the power and ground sides of a circuit. All three ammeters indicate the same amount of amperage (current) is flowing throughout the circuit.*

are basically the same as for series circuits, but with a couple of notable exceptions.

The first exception is: voltage will be equal (the same) everywhere on the positive side of the circuit and will not be divided between load devices, like in a series circuit. This is because each load device has its own wire, or conductor, connecting it to the power source. The same is also true on the ground return side; each load device has its own ground return. As a result, the ground return side of each load device registers 0 volts because each individual load device uses up all original source voltage.

Figure 1-10 displays 0 volts present on the ground return side of each individual bulb (load device) because each bulb uses up all of the original power source voltage. The ground return side of each bulb operates just like the ground return side at the end of a series circuit. However, with a parallel circuit, it is helpful to think of each load device as a simple series circuit itself—with a separate power source, load device, and ground return. Consequently, as more load devices are added to a parallel circuit, the greater the amperage from the power source to the load devices needs

to be (because total resistance of the entire circuit decreases as overall amperage increases, as in Figure 1-5).

The second exception is: each additional load device in a parallel circuit lowers the total overall resistance of the circuit and increases amperage.

Figure 1-11 (on page 24) illustrates how the addition of more load devices to a parallel circuit causes the resistance of the entire circuit to decrease and the amperage to increase (a teeter-totter effect)—unlike what occurs in a series circuit where the resistance of each load device adds to the total resistance of the circuit.

To understand this concept better, consider the following analogy. Imagine there are 20 people in a room with only one exit door, and everyone has to exit through the door, thereby creating lots of resistance. The people represent amperage, the room is the circuit, and their effort to get out is the resistance. Because there is only one door from which to leave, people slowly get out because of the effect of their combined resistance. Now assume the same 20 people are in a room with 20 doors. Everyone is able to exit the room quickly because each leaves through a separate door. Because there are

22

more doors, there is less resistance in the flow of people. The extra doors in the room represent additional load devices in a parallel circuit. Resistance to electron flow is reduced because a parallel circuit has more doors (load devices) from which the electrons can exit. As stated before, when resistance decreases in a circuit, amperage increases; so, adding more load devices (doors) lowers total resistance and increases circuit amperage. Figure 1-11 (on page 24) illustrates how each individual circuit within a parallel circuit operates like a series circuit. Each has a power source, load device, and ground return.

The third exception is: a parallel circuit requires all the power for the entire circuit to come from the same power source (the battery). This means the total amperage used by all the load devices must pass through the 20-amp fuse. Since each of the 12-volt bulbs in the individual circuits uses 6 amps, the total combined amperage for all the bulbs

(3 bulbs x 6 amps = 18 amps) must pass through the 20-amp fuse. This isn't a problem because the fuse won't melt unless amperage exceeds 20 amps.

However, at the end of the circuit is an electric motor (another load device); it requires 12 amps to operate. As long as the motor switch is open, the circuit uses only 18 amps to light the bulbs and the fuse is adequate for this application. But if the motor switch is closed (the motor now connects to the parallel circuit), the amperage requirement for the entire circuit increases to 30 amps (18 amps from the bulbs + 12 amps from the motor = 30 amps). This amount exceeds the rating of the 20-amp fuse, so the fuse overheats and melts (in order to protect the circuit), thereby cutting off power to the entire circuit. For this circuit to work, the 20-amp fuse has to be replaced with a fuse with a rating of more than 30 amps.

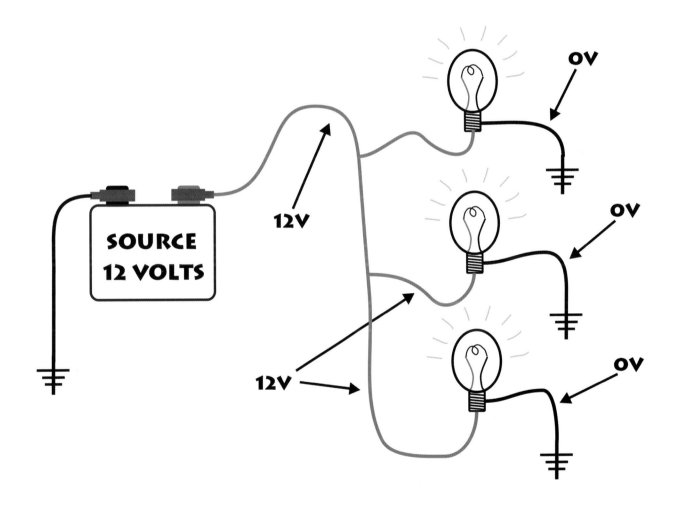

Fig 1-10. *Unlike series circuits, where the light bulbs (load devices) divide the total voltage between them, the bulbs in a parallel circuit are individually brighter because each bulb operates on 12 volts and uses all of the power source voltage.*

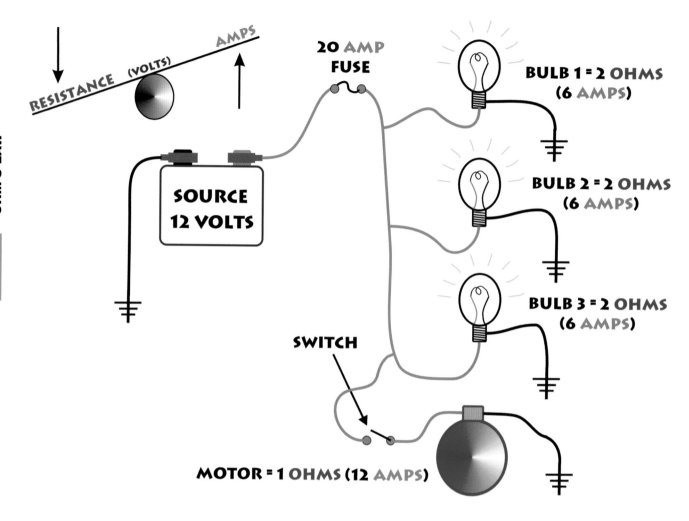

RESISTANCE (VOLTS)

AMPS

20 AMP FUSE

BULB 1 = 2 OHMS (6 AMPS)

BULB 2 = 2 OHMS (6 AMPS)

BULB 3 = 2 OHMS (6 AMPS)

SOURCE 12 VOLTS

SWITCH

MOTOR = 1 OHMS (12 AMPS)

Fig 1-11. *In a parallel circuit, each additional load device lowers the circuit's total number of ohms (resistance), thereby inversely raising the overall amperage flowing through the circuit. A higher-rated fuse and larger wires are required for this circuit to operate without melting wires or blowing the 20-amp fuse.*

Series/Parallel Circuits

Series/parallel circuits are rarely designed for automobiles, but they can be created whenever a poor connection is present in a parallel circuit. Remember, any form of resistance is considered a load device. Loose or corroded connections have resistance; therefore, whenever they are present, a series or series/parallel circuit is formed. The load devices in circuits with unwanted, high resistance operate with less voltage and amperage than intended; conse-

quently, all load devices connected to the circuit can be affected by the resulting reduced voltage and amperage.

Figure 1-12 displays a series/parallel circuit designed to operate a headlight and taillight. The addition of a light bulb between the battery and the other load devices represents a bad connection that is in series with the other load devices. The presence of the bad connection (bulb) adds resistance, lowering the available voltage and amperage necessary for the intended load devices

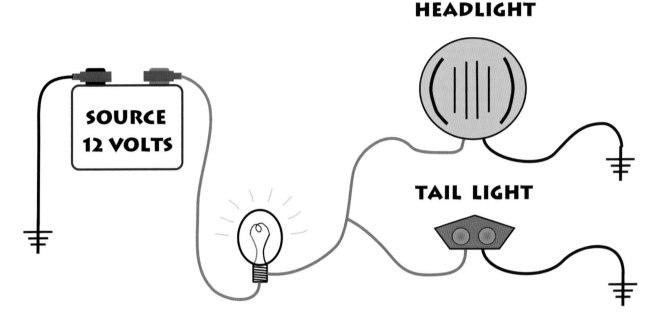

HEADLIGHT

SOURCE 12 VOLTS

TAIL LIGHT

THE BULB'S RESISTANCE REPRESENTS A POOR CONNECTION

Fig 1-12. *The presence of loose or corroded connectors (represented by the small light bulb) have turned this parallel circuit into a series/parallel circuit. The bad connection/bulb is in series with the headlight and taillight and divides the voltage between them. Thus, the originally intended load devices won't have enough voltage or amperage to operate properly.*

(headlight and taillight) to operate. Because the bulb uses some voltage from the power source, less than 12 volts are present in the wires going to the two lights. Since the headlights and taillights are designed to operate on 12 volts and not less, they won't work as intended and are consequently dim.

Replacing a headlight, taillight, or battery won't solve this problem either—but you'd be surprised how often this incorrect solution is attempted. The only way to repair what's really wrong is to find the bad connection; in this case, simply replacing the small bulb with a section of wire will eliminate the unwanted, high resistance so both the headlights and taillights will shine brightly again. The trick is to find the bad connections without unraveling wiring harnesses or removing parts, and that is what Chapter 2, Voltage Drop Testing, is all about.

CHAPTER 2
VOLTAGE DROP TESTING

WHAT VOLTAGE DROP TESTING DOES

Consider the following scenario: The owner (let's call him Bob) of a six-year-old car tries to start the engine on a cold morning. The starter cranks slowly a few times and then makes a clicking sound. A tow truck is called and after a jump, it starts right up. Bob takes the car to a local mechanic who claims he can fix the problem by replacing the battery. He does so, and the engine starts right up. Problem fixed? Maybe. Lucky for Bob, the car starts and runs okay every morning for a few weeks. Bob is happy the problem has been fixed. Several weeks later, though, the same thing happens again; the engine cranks too slowly. Bob, now angry, takes the car back to the mechanic and demands an explanation. He gets one. This time the mechanic says it has to be a faulty starter. Bob reluctantly replaces the starter, having lost confidence in the mechanic. The car works fine until several days later, Bob gets the same result again. Sound all too familiar? After spending $100 on

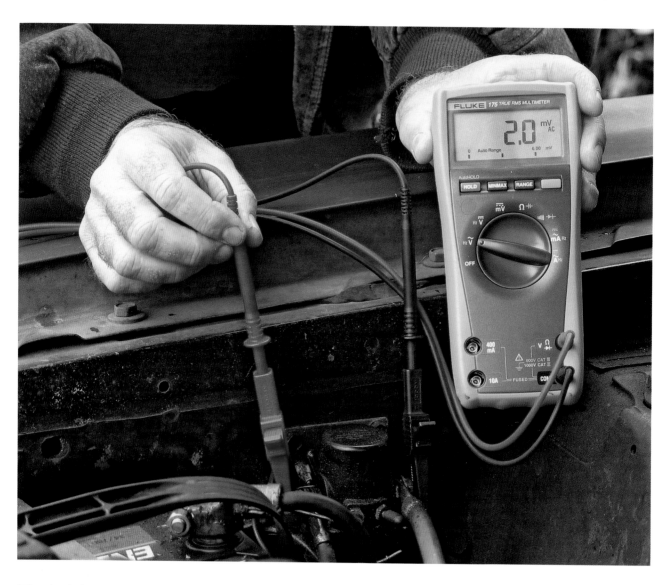

Voltage drop testing is the only reliable way to determine if a connection, or wire, can pass enough amperage to run a circuit. With the starter operating, the starter solenoid on this Ford truck is being tested for unwanted, high resistance using the voltage drop method. Courtesy Fluke Corporation

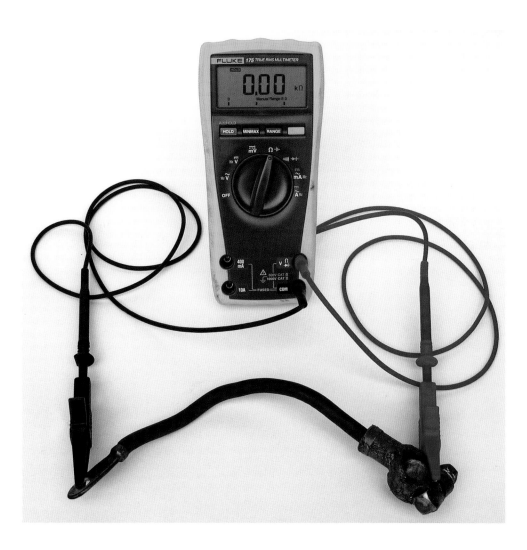

Even a good ohmmeter, like this Fluke 175, is only able to indicate that a battery cable is, essentially, not cut in half. While continuity can be shown on an ohmmeter, the battery cable could still have unwanted, high resistance that can only be measured by a voltage drop test. Courtesy Fluke Corporation

a new battery, $150 on a new starter and $175 for labor, not to mention all his wasted time and aggravation, the original problem is still there.

As it turns out, the cause of the slow-turning starter is a loose bolt that attaches the battery ground cable to the engine block. Poor electrical connections are the most common cause of many electrical problems. While this statement may seem obvious, in reality, few mechanics think to look first to the electrical connections as the source of an electrical-related problem with the car. In fact, this scenario can happen with almost any electrical component—alternators, heater-blower motors, headlights, windshield wipers, and others. Home mechanics, as well as many professional technicians, have the same difficulty diagnosing this type of problem.

Fortunately, there is a simple way to test for bad connections, switches, or wires as the likely cause of a problem with an electrical circuit—a method called voltage drop testing. It measures resistance within a circuit using a voltmeter (not an ohmmeter). Voltage drop testing is the easiest method to rule out the wires, connections, battery cables and terminals, relays, and switches as the causes of an operating problem with an electrical circuit. The major advantage of performing this test is that nothing needs to be disconnected in order to perform it—no unraveling of the wiring harness and no removing starters, alternators, or any other components.

Simply because a wire or electrical connection looks good doesn't mean it is good. The only way to determine if unwanted resistance is slowing down the flow of electrons passing through a connection, or along a wire, is to measure the drop in voltage within the circuit as it operates.

DON'T USE AN OHMMETER

Many shade tree mechanics, and even some professional technicians, insist that finding a bad wire, switch, or connection can be accomplished using an ohmmeter. While this is partly true, it is certainly not the best method (or even a good one!) for diagnosing loose connections and

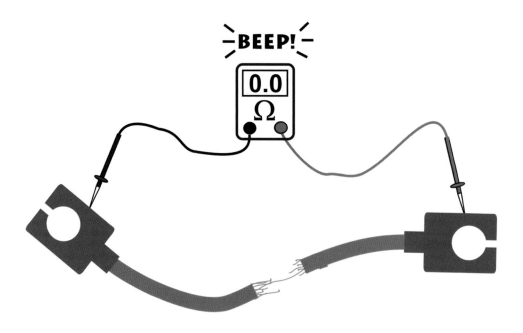

Fig 2-1. *High resistance equals low amperage flow in this battery cable. The ohmmeter registers continuity between the ends of the cable, but this battery cable still won't start an engine because of high resistance.*

other similar electrical resistance problems. Attempting to test a wire using an ohmmeter can reveal only one limited fact—whether the wire has continuity (a connection exists) between the two points being measured. However, it's the quality of the connection and/or wire that will determine if the circuit can carry enough amperage to operate.

When checking continuity using an ohmmeter, the meter's internal battery sends out only a few milliamps (thousandths of an amp) of current through the wire or connection, thereby enabling the meter to read in ohms. Battery and alternator cables and wires are capable of carrying hundreds of amps. Even heater-blower or windshield-wiper motor circuits may require 15 to 30 amps to operate. The few milliamps from an ohmmeter cannot simulate the operating conditions that the wire or connection being tested is subjected to by high amperage load devices.

Figure 2-1 illustrates the problem of using an ohmmeter to check a bad battery cable. While the ohmmeter registers continuity, the cable obviously does not have enough capacity to carry the current required to start an engine. Unwanted, high resistance in the battery cable will not show up when using an ohmmeter for testing.

Another problem with attempting to use an ohmmeter to measure a wire's resistance is its inherent inability to measure low resistance. Using an automotive ohmmeter to check the resistance of any wire or load device (such as a starter motor) will not work because resistance could be as low as 0.05 ohm. To accurately measure resistance less than 0.1 ohm requires a laboratory-grade ohmmeter, which costs thousands of dollars. Since resistance measurements in this low range are

also affected by temperature and humidity, an automotive ohmmeter simply does not have the required sensitivity.

The only reliable method for locating a poorly conducting wire or connector is to use a voltmeter to measure the voltage drop across the section of circuit being tested. Voltage drop testing works because even inexpensive digital voltmeters can accurately measure small amounts of voltage—as low as 1/1000 volts (1 millivolt). Voltage drop testing provides a comparison of voltage between two points in an operating circuit by displaying the difference or loss in voltage between them. The greater the unwanted, high resistance, the greater the voltage drop.

HOW A VOLTMETER MEASURES VOLTAGE

Because voltage drop testing requires the use of a voltmeter, understanding how a meter actually measures and displays voltage will help explain the results of a voltage drop test. A voltmeter displays a reading of 12 volts when connected to a 12-volt battery. This may seem obvious, but this reading actually represents the difference in electrical pressure between the positive and negative battery terminals. The red meter lead senses 12 volts; the black meter lead senses 0 volts. The computer chip inside the voltmeter simply subtracts the value of the black lead from the voltage level present on the red lead and displays the difference. Figure 2-2 (on page 29) exhibits the math occurring inside the voltmeter. An analog voltmeter (with a needle) uses the voltage difference between the red and black leads to power a small internal motor, which swings the needle across a numerical scale.

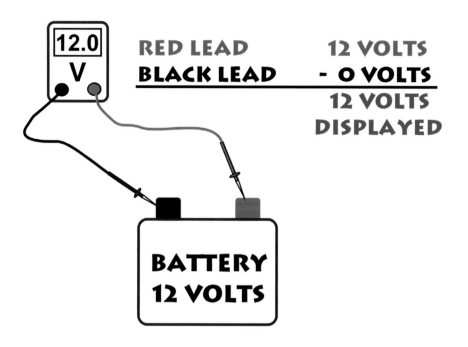

RED LEAD   12 VOLTS
BLACK LEAD   - 0 VOLTS
                12 VOLTS
               DISPLAYED

BATTERY
12 VOLTS

Fig 2-2. *A voltmeter measures the difference in electrical pressure between its leads. The value of the negative lead (0 volts) is subtracted from the positive lead (12 volts) to display the battery's total voltage—12 volts.*

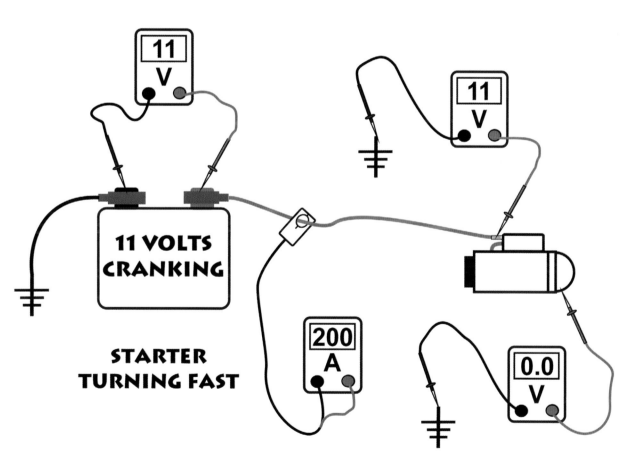

11 VOLTS
CRANKING

STARTER
TURNING FAST

Fig 2-3. *This starter circuit has low resistance; as shown indirectly on the ammeter, it carries enough current (200 amps) from the battery to start the engine. The voltmeters indicate no voltage is lost on either the positive or negative side of the starter circuit.*

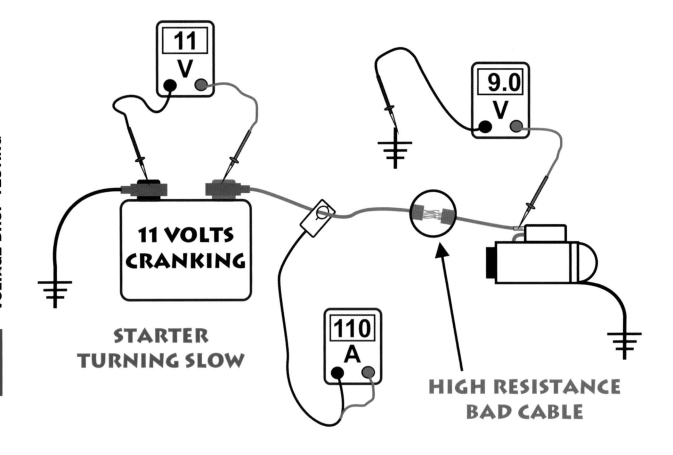

Fig 2-4. *A bad battery cable can prevent sufficient amperage from reaching the starter. As a result, the engine turns too slowly to start because the positive cable looses 2 volts between the connection at the battery and the starter motor.*

TESTING CIRCUITS DYNAMICALLY

Dynamic testing occurs when the circuit being tested is actually operating. This type of testing stresses all the components within a circuit and is more indicative of real-world operating conditions than other forms of testing. Current must be flowing through the circuit to measure voltage drop. Figure 2-3 illustrates the voltage present at various points in an operating starter circuit. With the starter cranking the engine, an ammeter reads 200 amps flowing throughout the starter circuit. Battery voltage normally drops to around 11 volts because of the amperage load from the starter. Voltage measured at the starter is also 11 volts. This demonstrates that the positive battery cable has low resistance and no voltage is lost between the positive battery terminal and the starter. The voltmeter reading for voltage drop on the ground return side of the starter circuit is 0 volts—this indicates the ground cable and all connections back to the battery have low resistance and no voltage is lost. The reason the ground side of the

circuit reads 0 is because the load device (the starter) uses up all available voltage.

Figure 2-4 shows what happens when high resistance (in this case, a bad battery cable) is present in a starter circuit. The starter cranks the engine too slowly for it to start. The positive battery terminal has 11 volts, but only 9 volts reach the starter motor—not enough electrical pressure to crank it fast enough to start the engine. The starter motor is designed to operate on a minimum of 11 volts, not 9 volts. High resistance in the positive battery cable causes a loss of 2 volts between the battery and starter. Because resistance is high in the circuit, amperage decreases and only 110 amps flow through the circuit (indicated by the ammeter) instead of the 200 amps the starter requires.

As mentioned above, an ohmmeter simply cannot accurately measure the high resistance present in the battery cable shown in Figure 2-4. Figure 2-5 illustrates why. To find the actual resistance present in the battery cable, both voltage and amperage must be measured. The math is

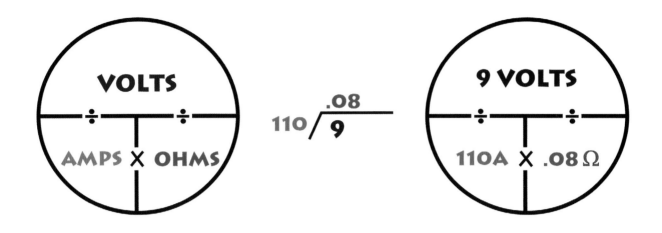

Fig 2-5. *With only 9 volts and 110 amps measured in this starter circuit, resistance is calculated as 0.08 ohm, far too low for an automotive type of ohmmeter to measure accurately.*

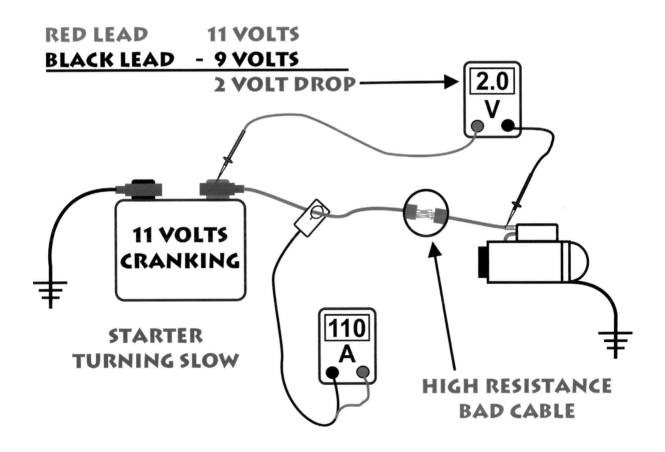

Fig 2-6. *When performing a voltage drop test on the positive side of a circuit, connect the positive lead to the source of highest voltage—the battery's positive terminal. Connect the negative lead to the point where the voltage is trying to get to—in this case, the starter battery terminal.*

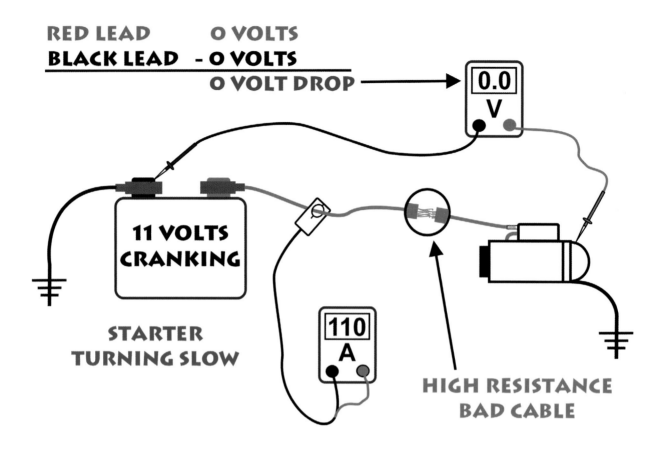

RED LEAD   O VOLTS
BLACK LEAD - O VOLTS
O VOLT DROP

0.0 V

11 VOLTS CRANKING

STARTER TURNING SLOW

110 A

HIGH RESISTANCE BAD CABLE

Fig 2-7. *The starter should use all the available voltage when it's operating. The starter's metal case is part of its ground path back to the battery's negative terminal. The voltage drop between the case and negative battery terminal should be 0 volts, indicating no voltage loss.*

simple: 110 amps divided into 9 volts equals a resistance of 0.08 ohm, or 8/100 ohm—far too low a value for an automotive ohmmeter to measure.

VOLTAGE DROP—POSITIVE SIDE

In the examples of voltage drop testing presented so far, it would be necessary to use several voltmeters to determine voltage drop in a starter circuit. However, only one voltmeter is needed when this test is performed. Figure 2-6 (on page 31) shows how the test is accomplished using only one voltmeter to measure the voltage drop on the positive side of the starter circuit. The red meter lead is connected to the power source, or the highest point of potential voltage (in this case the positive battery post), and the black meter lead is connected to the point where the voltage is trying to get to, which is the load device (starter's positive terminal).

The potential difference in voltage between the battery's positive terminal and the cable's connection at the starter is 2 volts, which is displayed on the meter. Remember, the voltmeter reads the value on the black lead, subtracts it from

the value of the red lead, and displays the difference. If, by mistake, the meter leads are connected backwards (red to the load device and black to the positive battery post), a digital voltmeter will display a negative number—this isn't a problem if you simply ignore the minus symbol when you read the display. However, if an analog meter is connected backwards, the needle will go to the far left, past zero, and no reading is possible. A simple rule to remember in order to connect voltmeter leads properly when performing a voltage drop test is: the red lead gets connected to the point where voltage is highest. This will always be the power source for the load device—usually the battery's positive terminal. The black lead gets connected to the lowest point of voltage, or where the source voltage is going to—the load device.

VOLTAGE DROP—GROUND SIDE

It's important to understand that once a voltage drop test has been performed on the positive side of a circuit, you're only half finished. Don't forget to voltage drop test the ground return side of the circuit! Checking for resistance

32

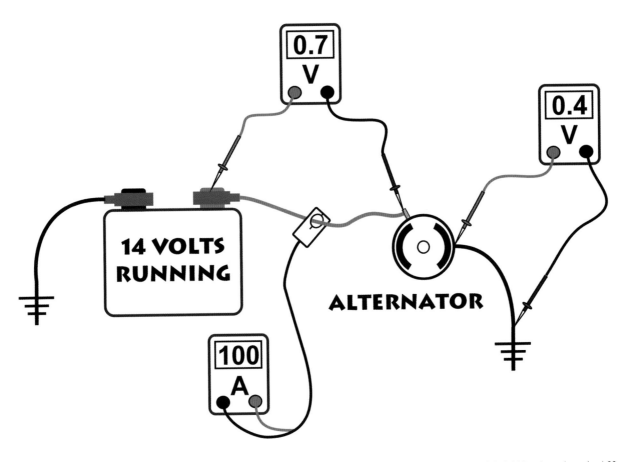

Fig 2-8 *Because of high alternator amperage output, voltage drop in late model alternator circuits is greater than on older models (which only produce about 60 amps). The ground side voltage drop of 0.4 is acceptable for this circuit.*

on the ground return is of equal importance as testing for it on the positive side. Remember the three rules of DC circuits? One of them states that: All available voltage in a circuit is used up by the load device. Although our example tests a starter circuit, this rule holds true for voltage drop testing of any circuit. If the ground return side of the starter circuit has low resistance, there should be no voltage lost on its ground return path. This path consists of the starter's metal case, negative battery cable, and the negative battery terminal. There should be a 0-volt difference between the starter's metal case and the negative battery terminal, thus indicating low resistance to current flow. The bad positive battery cable present in Figure 2-7 will have no effect on the ground side voltage drop test.

To measure voltage drop on the ground side of the starter circuit, connect the red lead to the point with the highest potential source of voltage—in this case, the metal housing of the starter motor. (When performing this test on other types of load devices, the red lead would be connected to the ground return wire.) The black lead is connected to the negative

battery post—the lowest point of voltage. Once the starter is cranked, the voltmeter will display the difference between the voltage potential on the red and black leads. In this example, the voltage drop on the ground side of the starter circuit is 0 volts. In other words, the ground return side has low resistance because no voltage is lost. In the real world, some loss of voltage will always occur between connectors and along wires or cables within a circuit because there is always some resistance to current flow, even in circuits that are operating properly. How much voltage drop is too much is always a matter of opinion.

HOW MUCH IS TOO MUCH?

There are two reasons voltage drop numbers vary. The first is that the greater the amperage in a circuit, the greater the voltage drop. For example, an alternator producing 40 amps will only have a voltage drop of 0.2 volt, assuming there is no unwanted resistance. Alternators found in late-model vehicles can produce over 100 amps and can have voltage drop numbers around 0.7 volts or higher. Other

The red lead is connected to the big wire on the back of the alternator. The black lead is connected to the positive battery post. With all electrical accessories on and the engine holding at 2,000 rpm, the voltage drop on the positive power side of the alternator circuit will be displayed.

factors affecting circuit voltage drop are wire length and/or size. For example, the positive wire between a vehicle's brake switch and the brake lights on a 30-foot trailer the vehicle is towing has a greater voltage drop than the same wire going only to the vehicle's brake lights. The longer length of wire connected from the vehicle to the trailer increases its resistance, thereby lowering the voltage present at the trailer's brake lights.

In general, voltage loss on the ground side of a circuit is almost always less than on the positive side. Ground return circuits often comprise the frame of a vehicle, engine block, or other large metal objects. These large vehicle components can be more simply thought of as really big wires, and as such, they will have low resistance to current

flow, causing a smaller voltage drop. The positive (power) side of a circuit will always be a wire or cable of some length and size and, therefore, will have greater resistance to current flow than the ground side of a circuit, thus creating a greater voltage drop.

The following is a ballpark guide to refer to when performing voltage drop testing on various types of automotive electrical circuits.

Starter Circuit, Maximum Voltage Drop
Positive Side, Small Starter/four-cylinder engine, 0.3 volts
Positive Side, Large Starter/V-8 engine, 0.5 volts
Negative Side, 0.4 volts
Starter Solenoids, 0.2 volts

Even if a battery cable connection looks okay, it still may have high resistance. Performing a voltage drop test while the starter motor is cranking the engine can expose a poor connection without the need for removing the battery cable.

Battery Terminals, 0.2 volts

Charging Circuits, Maximum Voltage Drop

Positive Side, 0.3 volts (alternator charging at 40 amps)

Positive Side, 0.7 volts (alternator charging at 100 amps)

Negative Side, 0.4 volts

Accessory Circuits (Headlights, Brake Lights, and Taillights), Maximum Voltage Drop

Positive Side, 0.2 volts

Negative Side, 0.2 volts

Computer Circuits (Ignition Modules and Fuel Injection Sensors), Maximum Voltage Drop

Positive Side, 0.1 volts (computer circuits are low amperage)

Negative Side, 0.06 volts (computer grounds are very sensitive to loss of voltage)

LOCATING THE BAD SPOT

In Figure 2-9 (on page 36) is an example of how a voltage drop test can be used to locate a bad switch in a circuit. To isolate a point of high resistance in this light circuit, a voltmeter is connected to the positive battery post and the light bulb. The voltage drop (resistance to current flow) of 0.5 volts is too high and causes the light to be dim. To determine if the switch or connector is at fault, a voltage drop test is performed across the switch and connector. The minimal drop in voltage of 0.1 volts at the connector indicates there is no excessive resistance to current flow— thus, the connector is good. However, the voltage measured across the switch is 0.4 volts—way too high. The switch is causing the light bulb to be dim. Simply cleaning the switch

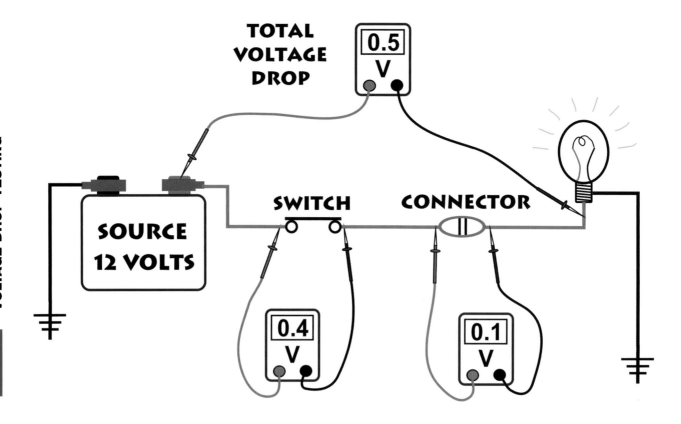

Fig 2-9. *By moving the voltmeter along the circuit, the point of high resistance can be located. The switch in this circuit has an unacceptable voltage drop of 0.4 volt. Time for a new switch!*

will lower its resistance and increase the voltage going into the bulb, thereby making it shine brightly again. This method of moving the voltmeter test leads along the circuit to locate the point of high resistance will work on any circuit.

Voltage drop testing is a good way to save a great deal of time and money in the quest to solve many electrical problems. The entire length of both positive and negative sides of a circuit can be checked without disconnecting any wires or connectors. For example, if the brake lights on a car are dim when the brakes are applied, it may only take two or three measurements to find the point of high resistance. To measure the voltage drop for the entire ground side circuit, connect the voltmeter's red lead to the ground wire for the brake light and the black lead to the negative battery post. (Don't forget to step on the brake pedal—the circuit must be operating during a voltage drop test!) If the voltage drop on the ground side is acceptable, high resistance will likely be found on the positive side of the circuit. Since a brake circuit uses a switch to control power to the brake light, performing a voltage drop test

across the switch is the most likely place to find high resistance. After connecting the red lead to the power source for the switch and the black lead to the other side of the switch, the resulting voltage drop will tell you if the switch needs to be replaced.

TEST, DON'T GUESS

Here are a few simple things to keep in mind when performing voltage drop testing: (1) Make sure the circuit being testing is operating—if no current is flowing in the circuit, there is no voltage drop to measure. (2) Test both the positive and negative sides of the circuit. If the voltage drop is too high on one side of the circuit, start isolating the point of high resistance by moving the test points along the circuit. (3) Most importantly, practice voltage drop testing on circuits that are working properly. That way, when you have to actually test a problem circuit, you'll know what's normal and what's not. (4) Finally, voltage drop testing is far easier than removing a starter or alternator, or replacing a battery. So, test, don't guess, before wasting time or money.

SECTION II
TOOLS

CHAPTER 3
ELECTRONIC TESTING TOOLS

Using a car's factory jack and lug wrench is okay for changing a flat tire, but the job is much easier using a hydraulic floor jack and 1/2-inch impact gun connected to an air compressor. Both sets of tools get the job done, but the latter is less work (and who likes extra work?). The same concept applies to automotive electricity. Using only a test light only gets limited results when trying to solve an electrical problem. Add a $10 voltmeter and diagnostic potential dramatically increases. Although friends or loved ones may complain about you owning every tool known to

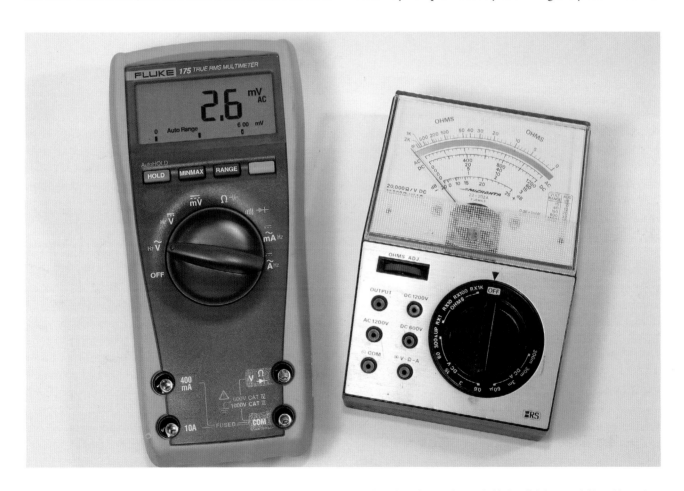

An analog meter (right) uses a needle to display readings. The meter's dial must be set to the right scale to read correctly. Modern digital meters (left) provide readings with electronically displayed numbers. Some, like this Fluke 175, have an analog bar graph that acts like an electronic needle—the best of both old and new.
Courtesy Fluke Corporation

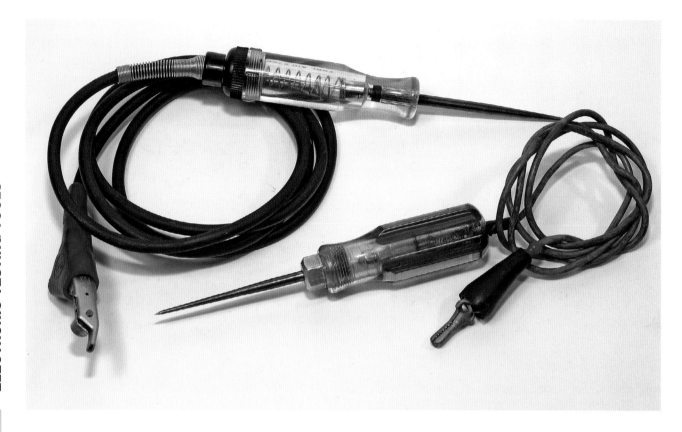

In addition to using heavier gauge wire, a quality test light (upper left) has a strain relief (metal spring or rubber) shield located where the wire exits the test light's body. This guard prevents the wire from breaking, even after years of use.

mankind (or wishing you did!), you can tell them you've only scratched the surface since you only have just enough tools necessary for the job at hand. When the next task comes along, you just may have to buy more. Like wrenches, sockets, and screwdrivers, you may never actually finish purchasing electronic test tools.

You may wonder if a $500 digital multimeter is really required to work on automotive electrical systems. The answer depends upon your diagnostic goals. If you're working on becoming a professional technician (or already are!) time is money . . . so an expensive digital multimeter (DVOM), with all its bells and whistles, will save you hours when diagnosing a problem. Home technicians usually don't have the same time constraints professionals do and can make do with less elaborate test equipment, so the purchase of an expensive multimeter may not be warranted. Even so, you should always buy quality tools. After working around cars for any length of time, people learn that the socket set bought at a local mass retailer is good enough for the toolbox in the trunk of a car or under the seat of a pickup truck, but for anything other than emergency repairs it's problematic.

Buying a quality digital multimeter and test light is a good starting point. Adding a logic probe to your electronic toolbox will also increase your ability to solve electronic ignition and onboard computer dilemmas. Let's start with the most basic tool—a 12-volt test light.

TEST LIGHTS

An automotive test light is a bulb mounted inside a plastic housing with a pointed probe at one end and a length of wire exiting the other end with an attached alligator clip. When the alligator clip is attached to ground, the pointy end of the test light can be used to probe wires or fuses to check for the presence of voltage. Once the probe of a grounded test light (alligator clip attached to ground return) touches a power wire, the internal bulb lights up.

Automotive test lights have been around a long time and are great tools for testing for the presence of voltage. However, test lights will not indicate levels of voltage. In fact, anything below 8 volts may not even make them light up. They are also useful for activating relays and solenoids. The probe of a grounded test light can be touched to the trigger of the terminal/wire of a relay to energize it. A grounded

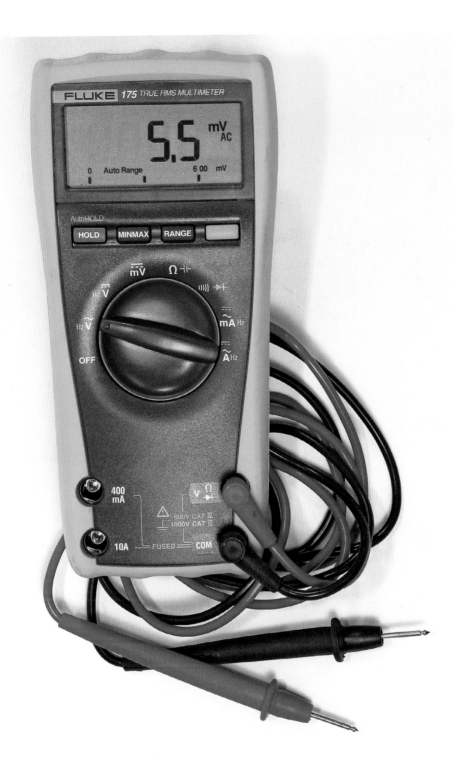

In addition to reading voltage and ohms, the Fluke 175 is a true RMS–type multimeter that can accurately read frequency produced by automotive-type AC pickup coils. Courtesy of Fluke Corporation

jumper wire will do the same thing, but a test light is safer. If you touch the probe to the wrong wire/terminal at the relay, the test light will just light up instead of causing a spark that could possibly damage the circuit.

Test lights can also be used in place of an AC pick-up coil, hall-effect switch, or optical distributor trigger to send a signal to an ignition module. In essence, the test light fools the ignition module into reacting as if it is receiving a signal from an engine speed sensor. If the module is working, it will produce a spark from the coil—this is known as a "tap test." A test light can also be used on an ignition coil to check whether an ignition system's primary circuit is switching on and off properly, or for an injector pulse from the vehicle's computer. (These tests are covered in later chapters.)

However, not all test lights work in all of these tests because some have bulbs with too much resistance—not enough current can pass through them to provide a trigger for ignition modules or relays. Test light bulbs should have less than 10 ohms of resistance to trigger relays and perform ignition tap tests. If the test light being used has more resistance, simply replace the bulb with one of higher wattage (lower resistance). Ironically, many of the more expensive test lights by Snap-On, Mac, and Matco use high resistance bulbs.

Test lights range in price from as little as $5 to over $25. So, what's the real difference between them? The more expensive ones use a quality wire for the lead and a better alligator clip. In addition, a metal spring or rubber strain relief guard is provided where the wire exits the plastic body of the test light, preventing it from breaking, even after hard use. Most importantly, quality test lights use hardened steel probes that don't require frequent sharpening and strong, durable plastic housings. Cheap test lights may break if dropped or banged about under the hood—so by the time you have bought several replacements, the cost of a good one would have been far exceeded. While both inexpensive and quality test lights do the same job with the same degree of accuracy, you'll get more life out of a quality one; so, save up and spring for the better quality. It's worth it in the long run (or handle your cheaper test lights carefully!).

Always test your test light before using it to check a circuit. Connect the test light to a good ground and touch the pointy end to something you know is hot—such as a positive battery terminal or fuse. Remember, test bulbs do eventually burn out. If you don't check your test light once in a while, you'll go crazy wondering why everything you test doesn't seem to have any power.

MULTIMETERS

The term voltmeter is practically a misnomer today since it's almost impossible to find a meter that reads only voltage, unless you're looking in an antique shop. The correct term for voltmeter is digital multimeter, or digital voltmeter. If you're serious about diagnosing and/or repairing electrical problems in your car or truck, a good digital voltmeter is as basic and necessary a tool as a 3/8-inch drive ratchet and socket set. Many amateurs, and even some professional technicians, try to find their way around a problem circuit using only a test light. Depending upon the type of information being sought, this method has its place; but more often than not, knowing only that voltage is present, but not how much, is simply not enough information to figure out what's wrong with a problem circuit. A good multimeter is essential if you don't want to waste a lot of time.

Over the last 20 years voltmeters have followed the pocket calculator, and most other electronic technology, by continually dropping in price and increasing the number of extra features available. Multimeters come in two basic flavors: analog and digital.

Old analog meters use a needle and fixed numbered scale to display readings. Analog meters are difficult to read accurately because it's hard to line up the needle with the numbers on the meter's face. In addition, if an analog meter is dropped, the mechanical movement of the needle can be damaged, producing inaccurate readings of which the user may be unaware. Nowadays, you have to look hard to find a new analog multimeter (again, try the antique shop) since most have been replaced by digital electronics.

A DVOM offers several advantages over analog meters. It displays electronic numbers via a liquid crystal display (LCD). Even if the meter gets banged about a bit, as long as the display is working, the meter's most likely still okay. Another major difference between the two types of meters is their internal resistance—an analog meter has low internal resistance (usually only around 100,000 ohms), while a digital meter typically has 10 megohms (10,000,000 ohms). Consequently, an analog meter will give false readings if used to measure high resistance computer circuits and can even cause damage to some automotive electrical components.

Voltmeters, multimeters, and DVOMs are all capable of measuring a variety of electrical inputs. All read volts (both AC and DC) and ohms. Other test functions performed include amperage, frequency, continuity, diode, capacitance, and temperature. Additional features include auto shut-off, digital smoothing, internal fuse protection, display backlighting, and analog bar graph display.

You can spend as little as $10 or as much as $500 for an automotive-type DVOM. Just as a $50 car stereo plays the same music as a $2,000 stereo, all digital voltmeters read electrical values, but all are not the same. The one that's best for you depends on the level of automotive diagnostics you want to pursue. Spending more money on a quality DVOM buys you durability, ease-of-use, extra features (oftentimes the only way to figure out a problem), quality meter leads, availability of accessories and product support.

AUTO-RANGE FEATURE

The majority of DVOMs sold today have an auto-ranging feature that senses volt or ohm input levels and automatically displays correct readings, including the rounding of numbers and correct placement of a decimal point. Using a meter without this feature requires a user to choose the correct scale before a meter reading can be interpreted. For

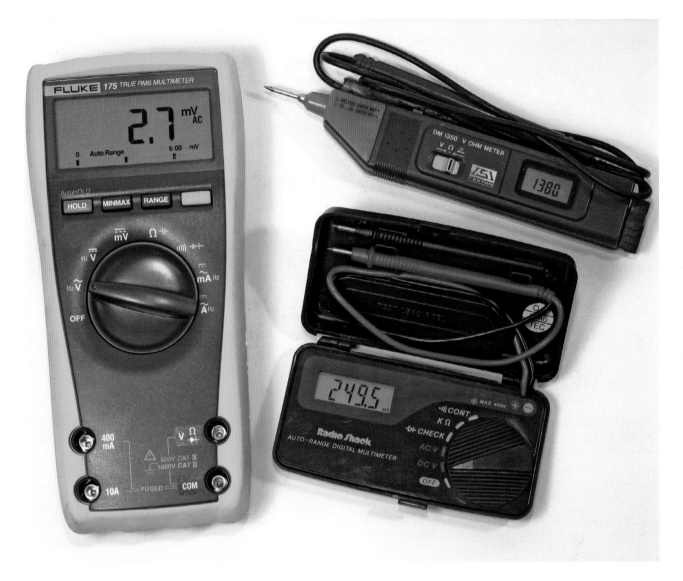

The Fluke 175 digital multimeter (left) has a bar graph feature not found on less expensive digital meters. Radio Shack's pocket digital multimeter (lower right) has more functions and gives more bang for your buck than other meters costing several hundred dollars a few years ago. Courtesy of Fluke Corporation

example, to read voltage from a 12-volt battery on a DVOM without auto-range, the 0–20-volt scale must be selected, thus allowing the meter to read between 0 and 20 volts.

Measuring resistance can be more difficult than reading voltage when using either an auto- or nonauto-ranging meter, since resistance values can cover a wide range of 1 to more than 1 million ohms. There are a few simple rules to help interpret ohm readings on both meter types. We'll use the resistance of an ignition coil to demonstrate how to do this.

An ignition coil's secondary windings have a resistance of 11,800 ohms. Using a DVOM with an auto-ranging feature, connect the meter leads to the coil's secondary terminals and set the function switch to ohms; a reading

should display. The meter's auto-ranging feature senses the coil's resistance and displays the correct reading—11.80k ohms. The "k" (thousand) symbol means you should add three zeros to the right of the decimal, making the reading 11,800 ohms—not 1,180 ohms; if the "k" symbol is not displayed, the reading would be complete—1,180 ohms. If the coil's secondary windings are open, the reading would be 11.80m ohms. The "m" (million) symbol replacing the "k" symbol means you move the decimal point six places to the right (the display does not have enough room to show this). The reading is now 11 million ohms, or 11,800,000 ohms. These same rules apply to nonauto-ranging meters. When using the 2k scale to measure ohms, the readings are

Before reading voltage or resistance, this Yuasa digital meter without an auto-ranging feature must be set to the correct scale. Courtesy Yuasa Battery

between 0 and 2,000 ohms. By selecting the 20k scale, the meter will read between 0 and 20,000 ohms; similarly, the 200k scale reads between 0 and 200,000 ohms.

FLUKE MULTIMETER FEATURES

The Fluke corporation has been manufacturing digital meters for the automotive market for years. Its Series 170 is a good example of an automotive digital multimeter available in a mid-range price (around $190).

There are two basic designs of voltmeters—averaging and true RMS (root mean square). Both meters measure voltage and ohms, but the RMS model more accurately assesses distorted sine waves. AC pickup coils and other AC

voltage–producing speed sensors have lopsided waveforms (sine waves). An averaging type of meter may not be able to read the output from a distorted sine wave. The Fluke 175 is a true RMS type of meter that can accurately read the frequency produced by AC automotive pickup coils.

Another difference between DVOMs is an analog bar graph. Meters with this feature provide both digital and analog displays. The bar graph acts the same as an electronic needle, only much faster. On a Fluke 175, the bar graph updates about 40 times per second—10 times faster than a digital display—allowing a user to detect changes in readings that occur too quickly for meters not equipped with bar graph features to display. This feature is also useful when checking for a bad throttle position sensor (TPS) signal or when watching the performance of a speed sensor with an intermittent problem—both tests are only possible using a meter with an analog bar graph display.

A Fluke Model 175 Digital Multimeter is a good example of a meter loaded with various features and functions, which can be accessed via a rotary switch and numerous buttons found on the meter face. Other manufacturers offer similar bells and whistles. To show the range of functions a modern digital multimeter is capable of performing, what follows is a description of the various features found on a Fluke 175 Digital Multimeter.

Rotary Dial Positions

The rotary dial on the meter face is rotated to select various tests and functions. Rotary switch selections are identified by yellow and white letters or symbols. All yellow functions are accessed via the yellow function button located on the top-right of the meter face just under the display. Pressing this function button twice returns the meter to the tests and functions indicated by white symbols or letters—these are accessed by simply turning the rotary switch. Here are the common rotary switch positions and their corresponding functions:

1. **Off:** Switches the meter off. If no input readings change for 20 minutes, the meter automatically enters into sleep mode, thereby saving its battery. This feature can be disabled for recording voltage over time.

2. **AC Volts/Hertz:** Measures AC voltage between 0.1 millivolt (mV) and 1,000 volts. Pressing the yellow function button switches the meter to AC capabilities, enabling it to read AC voltage frequency or hertz. AC voltage frequency testing is useful when checking AC pickup coil types of sensors.

3. **DC Volts/Hertz:** Measures DC voltage between 1 millivolt and 1,000 volts. Pressing the yellow function

button switches the meter to DC mode, enabling it to read DC voltage frequency or hertz. DC voltage readings are the most commonly used function on any meter. DC voltage frequency testing is useful when checking sensors with digital outputs, like Ford manifold absolute pressure (MAP) sensors or General Motors (GM) mass airflow (MAF) sensors.

4. **DC Millivolts:** Measures DC voltage between 0.1 and 600 millivolts. Performs essentially all of the same functions as the one mentioned previously, except this function is useful for measuring voltages on a smaller scale (usually measurements less than 0.6 volts).

5. **Ohms/Farads:** Measures resistance in ohms between 0.1 and 50 million ohms. The ohms function is useful for measuring resistance of sensors, relays, switches, plug wires, and other components. Pressing the yellow function button switches the meter to farad mode, enabling it to read in farads. This function is used to measure capacitance or capacitors; however, capacitance function is seldom used in automotive applications.

6. **Beeper/Diode:** Beeps whenever the meter leads touch each other, this being the test for continuity. Continuity testing is useful for determining if a wire is broken. Pressing the yellow function button switches the meter to test diodes like those used in alternators, or spike diodes used in other automotive circuits.

7. **AC/DC Milliamps:** Measures AC/DC milliamps from 0.01 to 600. Used to measure small current draw and for checking parasitic amperage draw from onboard computers, radios, clocks, and other devices. Pressing the yellow function button switches the meter to read AC frequency from 2 hertz to 99k hertz.

In addition to reading volts, amps, ohms, and frequency, this Fluke Series 170 multimeter has "Hold" and "MinMax" features. The yellow button switches the meter's reading modes between white and yellow lettered dial functions. Courtesy of Fluke Corporation

8. **AC/DC Amps/Hertz:** Measures AC/DC amps from 0.01 to 10. The amp function is protected by an internal fuse; it's useful for checking circuits up to 10 amps.

Pressing the yellow function button switches the meter to read AC frequency up to 99k hertz.

9. **Hold Button:** Holds or freezes the display when pressed. Pressing the hold button twice switches the display to auto hold. In this mode, the meter holds the reading until it detects a new stable reading; then the meter beeps and displays the new reading. This feature is useful when you can't see the meter (like when you are working under the dash) but want to know if you've probed a wire with power or other signal on it.

10. **MinMax Button:** Pressing the MinMax button lets the meter record minimum, maximum, and average values over time. By pressing the button once, the meter takes three stepped readings over time. Holding the button for one second ends the MinMax recording. The meter's sleep

mode is disabled in MinMax record mode. This function is useful for reading changes over time when you can't keep an eye on the display, and works really well when trying to figure out intermittent electrical problems.

11. **Range Button:** Powers up in auto-range mode. Pressing the range button changes the meter to manual range, thus allowing you to change where the decimal is located. Pressing the button moves the decimal point one place in the display. This feature is useful when you only need to know large changes in voltage or resistance.

12. **Yellow Button:** Switches the meter between different modes. This button also disables sleep mode if pressed when the meter first powers up.

Terminal Jacks

The Fluke 175 meter has four terminal jacks. The common or ground (COM) jack serves as the ground return for all measurements. The volt/ohm/diode jack (upper right) is for

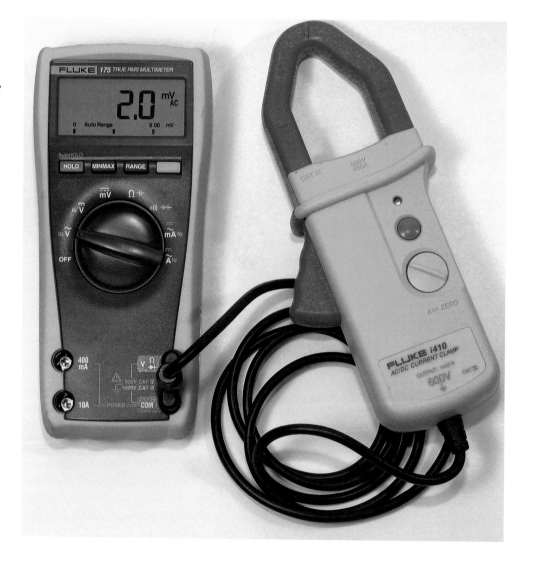

The Fluke i410 AC/DC current clamp can be used with any digital multimeter capable of reading in millivolts. This current clamp will measure up to 400 DC amps and is ideal for starting and charging systems diagnosis. Courtesy of Fluke Corporation

This Radio Shack logic probe can detect signals from AC pickup coils and hall-effect switches. It can also sense a fuel injector pulse from the vehicle's electronic control module (ECM).

testing voltage, continuity, resistance diodes, capacitance, and frequency measurements. Both jacks on the left are for measuring amps; each is protected by an internal fuse that can be tested by using the meter leads. By touching the leads together, the fuses can be checked without taking off the meter cover. Also, the meter has a built-in safety mechanism that warns the user against inadvertently connecting the leads incorrectly. For example, if the leads are plugged into the jack for voltage testing but the meter is accidentally switched to read in amps (or vice versa), the word "lead" is displayed. This warns the user to move the meter leads to one of the amp jacks to prevent melting the fuse.

MEASURING AMPERAGE

There are two types of ammeters used to measure current—series and inductive. Some digital voltmeters are designed with built-in series ammeters that can measure as much as 20 amps (a 10-amp capacity is more common). A series ammeter is useful when trying to find a parasitic amp draw, as it can be connected in series with a circuit that is causing

a drain on the battery. The Fluke 175 has two ranges for its series ammeter—400 milliamp and 10 amps. The 400-milliamp range can be used to monitor small amounts of current commonly used by onboard computers, clocks, and radio memory; the 10-amp range can help locate a stuck relay or interior trunk light. Both ranges are protected by internal fuses.

For measuring starter motor or alternator output amperage, an inductive current probe should be used. Current probes are self-powered (usually by a 9-volt battery) and produce a millivolt output. As current flows through the starter cable or alternator output wire, a magnetic field forms around the wire. The inductive amp probe measures this field and converts it to a millivolt signal, which can be read directly on a voltmeter (1 millivolt equals 1 amp). The Fluke i410 AC/DC clamp can measure AC or DC amps from 1 to 400. The clamp will plug into any voltmeter that accepts a banana-plug jack style. Be careful when considering a purchase of an inductive amp clamp; most only measure AC amperage, which is not

Snap-On and other manufacturers market a red and green test light. This product is simply a heavy-duty version of a logic probe.

useful for automotive use. The Fluke i410 costs about $155 and is really worth the money if you anticipate diagnosing starter or alternator problems.

LOGIC PROBES

One of the most useful and inexpensive electronic test tools for automotive use is a logic probe. Exclusively used for electronic board testing, the logic probe has finally made its way into automotive use over the last several years. These handy little tools are sometimes referred to as "red and green" test lights. A logic probe can only sense two things—high or low voltage. However, its real value is its ability to detect electronic pulses. The outputs from the following components can all be detected (though not measured) using a logic probe: AC pickup coil speed sensors, digital square wave sensors (like Ford MAP or GM MAF sensors), optical distributor outputs, and fuel injector pulses from a vehicle's computer. Because of its high internal resistance, a logic probe is a safe tool to use on any sensor or computer-generated output.

A logic probe is powered by connecting the probe's red lead to any 12-volt power source and the black lead to any ground return on the circuit being tested in a vehicle. The pointy end of the probe is used to detect pulse outputs from

various sensors. A green light emitting diode (LED) indicates whether input voltage levels are below the threshold level of 0.8 volts—a red LED lights up if voltage is above the high threshold of 2.2 volts. If a pulse is present at the probe tip, the LED will flash or flicker.

Radio Shack's logic probe (part number 22-303) costs about $20 and has an orange LED that flashes if a pulse is detected. In addition, it has a beeper that changes tone depending on voltage level. Searching under "logic probe" or "red and green test light" on the Internet will yield other brands that suit your needs as well. They all do about the same thing, and all are invaluable when working on electronic ignition and fuel injection systems.

NOID LIGHTS

A noid light is a specialty tool, specifically designed to check for the presence of a fuel injector pulse at the injector wiring harness. A noid light plugs into the wiring harness in place of a fuel injector. When the engine is cranked over, the light flashes or flickers if a pulse is received. A set of noid lights can be purchased on the Internet or at most car parts stores for about $20. Most sets typically come with a number of different lights and connectors to accommodate various sizes of major manufacturers' fuel injector connectors.

However, as mentioned earlier, a test light can also perform this same function and is equally safe to use on all electronic fuel injection (EFI) systems. However, some EFI systems use a series resistor with the injector, and these systems may not pass enough current to light the test light. Noid lights usually require less voltage/current to light up and may work better in this respect.

SHORT FINDERS

If the same fuse keeps blowing, it's easy enough to look at a wiring diagram to determine the accessories powered by that circuit. However, it's altogether another matter to actually find the section of wire or connector that has shorted to ground. Service manuals rarely provide wiring harness routing locations, so oftentimes, the only way to find a bad section of wire is by taking lots of things apart—dashboards, engine wiring harnesses, head liners, and so on.

The task of finding a short in a circuit is much easier when you use a simple "short finder," which can be made at home using a resettable circuit breaker and compass. (Self-setting circuit breakers can be purchased at most auto parts stores, while sporting goods' dealers can supply a compass.) The circuit breaker connects directly to the fuse box and takes the place of the blown fuse. The bimetallic strip inside the circuit breaker heats and cools as current passes through it, causing the circuit to turn on and off. Each time the circuit receives power, a magnetic pulse is generated in the wire. Placing the compass needle near the wire causes it to deflect each time the breaker resets. By moving the compass along a wiring harness, the location of the short can be determined by watching the needle movement in the compass. It is always about the same strength until it is placed just beyond the shorted wire. As the compass needle moves less and less, you are moving

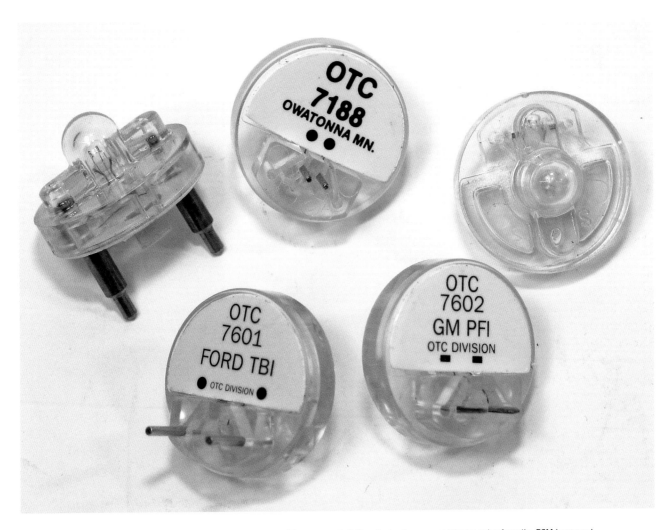

Each of these Noid lights fit different fuel-injection wiring harnesses. When connected, they flash whenever an injector pulse from the ECM is present.
Courtesy of SPX/OTC

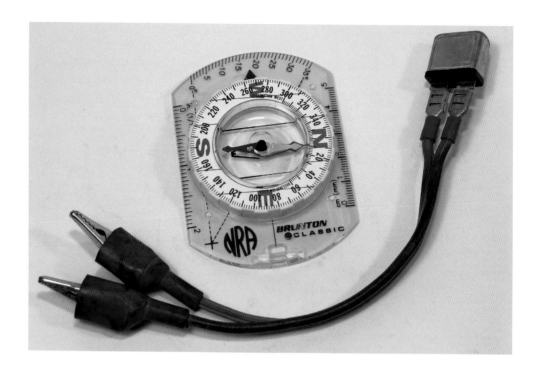

This homemade short finder consists of a 5-amp resetting circuit breaker, wire leads with alligator clips, and a compass.

This homemade short finder uses either a 5- or 10-amp resetting circuit breaker. In addition, it flashes a light and beeps. Other functions include a 20-amp breaker for testing wiring harnesses or powering up a fuel pump.

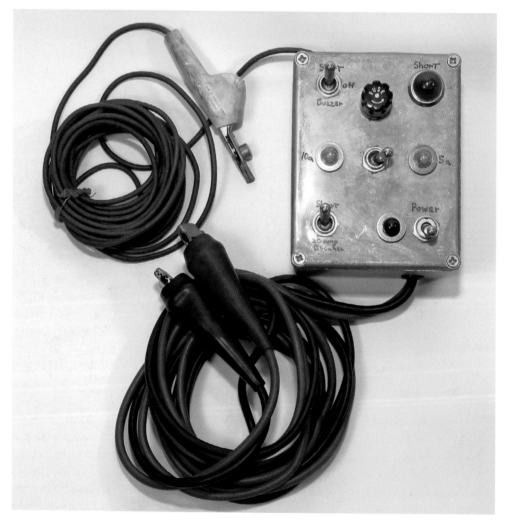

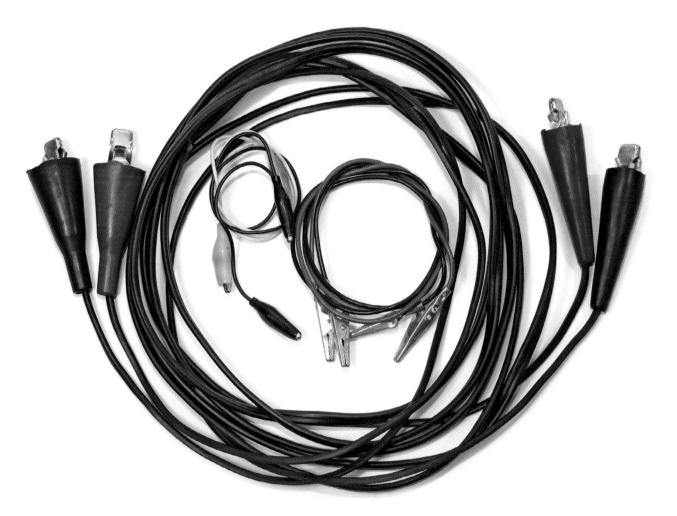

One can never have too many jumper wires. They come in all different sizes, as shown here. The large set of jumpers was homemade using 14-gauge lamp cord; it comes in mighty handy for getting power from the front to the back of a vehicle or trailer.

farther away from the shorted wire. With a little practice, you should be able to narrow down the location of the shorted wire to within 6 inches. Also, the compass can sense a magnetic pulse through sheet metal, upholstery, and just about everything else. You can either make your own short finder (as described above) or purchase one online by searching the Internet under "automotive short finder;" the cost should be around $20.

JUMPER WIRES AND ACCESSORIES

Jumper wires are basically nothing more than various lengths of wire with alligator clips connected to their ends. Jumper wires have numerous uses and a serious technician always has at least two (red and black) clipped to the outside of a toolbox. The best jumper wires can be easily made from silicone wire and good quality alligator clips. Heavy lamp cord (14 gauge) with larger clips also works well when you want the convenience of having both power and ground sources at the rear of a vehicle in order to test fuel pumps or when working on trailer lights. If you don't want to make your own, packages of jumper wires in most sizes can be purchased at Radio Shack or via the Internet.

Measuring voltage or other outputs on wires that disappear into a connector or sensor can prove problematic. Picking up a knife or wire stripper to scrape off wire insulation is not a good way to tap into a wire since it could cause damage to the wire. A better method for testing connector or sensor wires is to use a seamstress dress pin (available at most fabric stores). The dress pin can be inserted between a wire and connector until contact with

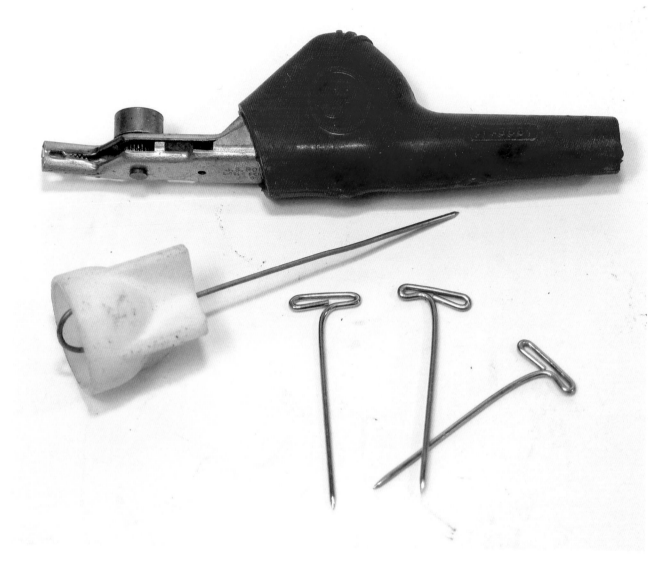

A JS Popper alligator clip (top) equipped with a bed-of-nails. This tool can pierce wire insulation and provide a place to connect a voltmeter. The T-shaped pins (bottom) are available from any fabric store and work well to back probe connectors.

the terminal inside the connector is made. The T-shaped end also provides a place to attach meter leads. Furthermore, when the pin is removed, there's no damage to the wire or connector.

The best tool around for tapping into a wire with no connector is a JS Popper alligator clip by Mueller Electric Company. This company's JP-8783 alligator clip is equipped with a bed-of-nails (sharp spikes clustered close together). These clips are designed to reach the conductor inside a wire by piercing the insulation without causing damage to the wire. They can attach directly to meter leads or jumper wires. Search for this specialty tool on the Internet under "JS Popper" or "Mueller Electric."

COIL TESTERS

Testing an ignition coil is oftentimes problematic. When using an ohmmeter to measure the resistance of the coil's primary and secondary windings, does it mean the coil is okay if the resistance of the windings is within specification? Not necessarily. Sometimes an ignition coil with internal resistance that meets specifications won't produce a spark when connected to the vehicle's ignition system, even though the windings check out okay with an ohmmeter. A better way to test a coil is to use a universal ignition coil tester, which can be made simply with a condenser and some jumper wires. A coil tester basically functions like a set of ignition points and can be used to fire any type of

An old test light body, condenser, wire, and some alligator clips can be combined to create a homemade coil tester. If a coil is good, then this tool will make it produce a spark.

ignition coil. When connected to an ignition coil with battery power, the coil tester will cause the coil to produce a spark if it's good. This dynamic test is truly effective for discovering a bad coil winding that could break down under an electrical load. A more complete description of how a coil tester works can be found in Chapter 6 on ignition systems.

SPARK TESTERS

Using a jumper wire held close to ground as a method of testing for ignition coil spark output will work, but it can zap you if you're not careful. A high energy ignition (HEI) tester is a safer device to use. It is basically a spark plug with no ground electrode and with an alligator clip welded to it.

About 25,000 volts are needed from the ignition coil to produce a spark at the tester, indicating a good coil that should be capable of producing enough energy for the ignition system. Consequently, this tester may not work on some points-type ignition systems because their secondary voltage output is too low.

An adjustable spark tester will also work and will provide an approximation of spark output in thousands of volts. This tester has an electrode that can be adjusted to increase the air gap the spark has to jump. A printed scale on the tester indicates how much voltage is required to jump the resulting air gap. Both testers work equally well, sell for less than $8, and can be found at most auto parts stores.

By observing the distance a spark jumps on the spark tester on the left, you can get some approximation of spark-plug-firing voltage. The air gap's width can be set by the user to increase or decrease the firing voltage required to jump the air gap. The HEI spark tester on the right has a fixed air gap that requires about 25,000 volts from the ignition coil to jump its air gap. This is a good indication of coil strength on most electronic ignition systems.

Any inductive timing light can also function as a poor man's scope. When connected to an ignition wire, the flashing light will skip a flash whenever a misfire occurs.

IGNITION TIMING LIGHTS

In addition to adjusting distributor ignition timing, an inductive timing light can also be useful for detecting ignition misfires. By installing the inductive clamp on the coil wire and then pointing it at your face, you can actually see misfires because the light skips a flash whenever a misfire occurs. Thus, the human eye and timing light serve to create a poor man's ignition scope. This technique also works on distributor-less ignition systems (DIS) as well. A timing light can also be used to detect an ECM that is not firing a fuel injector. By clamping the inductive probe around one of the wires going to a fuel injector, injector misfires can be observed in a similar manner. By connecting a timing light to a throttle body injector wire and pointing it at the injector, you can actually see the injector spray pattern when the light flashes in time with the injector pulse. Any inductive-type timing light will work for these tests. Timing lights are available at most auto parts stores and online.

BATTERY TESTERS

Using only an ammeter and starter motor to test a battery is a time-consuming task. There are a number of hand-held automotive battery testers on the market that make this job a whole lot easier and faster. The Owatonna Tool Company (OTC) makes several types of battery testers. Its Model 3180 battery tester can load a battery to 100 amps. The tester consists of a voltmeter, relay, and heavy-duty resistor. After connecting the tester cables to a battery, turn the load switch on for 10 seconds. Before releasing the switch, read the voltage display to determine if the battery is good or bad. The battery has to be at least 75 percent charged in order for all load-type testing to produce reliable results. OTC's Model 3181 battery tester draws 130 amps and has clamps that work on side-post battery terminals. Both OTC testers can test 6-volt or 12-volt batteries and have starter and charging tests as well. Hand-held battery load testers can be found at most auto parts stores and online.

If you want something a little more high-tech, OTC also makes a digital battery tester (part number 3191), which uses conductance technology to determine a battery's condition. These testers indirectly measure the available plate surface area of a battery needed to produce the chemical reaction that creates current. Thus, conductance testing provides a reliable indication of the overall battery state-of-health (how much life is left) and has a direct correlation to a battery's capacity to start an engine. Conductance testing can also be used to detect bad cells, shorted plates, and open circuits within a battery. Furthermore, unlike load-type battery testers, digital testers can also test a discharged battery.

An OTC 3180 Stinger battery tester pulls a 100-amp load during battery testing. Hand-held battery testers are more convenient than using the engine's starter motor. Either a 6-volt or 12-volt battery can be tested. Courtesy of SPX/OTC

Testing takes only a few seconds and the digital display includes actual cold cranking amps (CCA), open circuit voltage, and battery health. Unfortunately, these testers don't come cheap! Because of their price ($350 and up), these testers are mostly used only by professional technicians.

An OTC 3181 battery tester is more heavy-duty than an OTC 3180. It uses a 130-amp load to test a battery. The clamps work on both standard- and side-post battery terminals. Courtesy of SPX/OTC

LAB SCOPES

Oscilloscopes are commonly used in laboratory settings, but have also been used widely in the automotive industry since the 1950s. In the automotive repair industry, they were almost exclusively used for diagnosing ignition system problems until the late 1980s. At that time, newer digital lab scopes were introduced as tools that could "see" various inputs and outputs from electronic fuel injection systems. Computer sensors, including MAP, barometric pressure (BARO), AC pickup coils, hall-effect switches, optical distributors, and MAF all produce waveforms. A waveform is a shape derived from displaying voltage-over-time for a signal output. A lab scope can view and display sensor waveforms and, additionally, can measure and record waveforms from actuators, including those from fuel injectors, idle controllers, solenoids, and relays.

Why, you may wonder, should you use a lab scope when most (but not all) of the signals from the previously mentioned sensors and components can be measured with a multimeter? There are a few good reasons. For example,

a Ford MAP sensor produces a square wave that a multimeter displays in hertz frequency. As a MAP sensor's signal changes, the number of hertz increases or decreases; however, if the sensor has an intermittent problem, a multimeter will not be able to detect it. In some instances, use of a lab scope is the only way to actually see the square wave break down. This is most useful in those instances when the quality of the signal is of greater importance than the presence of the signal.

Another good example of a lab scope as the preferred tool is when measuring fuel-injector pulses from the ECM. A flashing noid light, test light, or logic probe connected to an injector harness will only confirm a signal exists; but again, the quality of the signal cannot be determined by these test methods. By contrast, a lab scope not only verifies an injector signal is present, but by analyzing the scope's digital display, you can determine if there is a problem with the vehicle's fuel injection computer. A lab scope can even detect a bad fuel pump motor or cooling fan motor before they quit working. By observing the waveform either motor produces, a bad

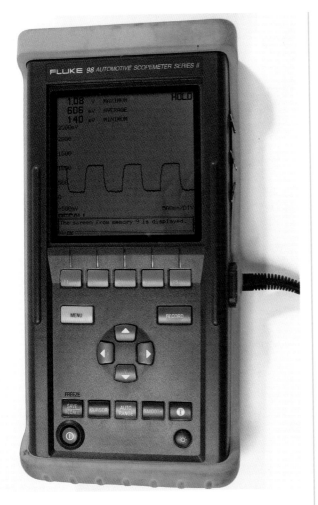

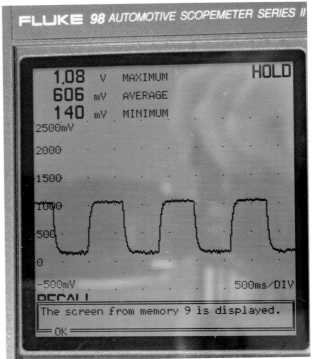

The ability to play back this recording of a hall-effect switch is sometimes the only way to "see" an electrical glitch that could cause an ignition misfire. Courtesy of Fluke Corporation

A Fluke 98 Scope Meter can record and play back both analog and digital electronic fuel-injection signals. It can also measure secondary ignition voltage and display spark plug firing patterns. Courtesy of Fluke Corporation

motor brush or armature winding can be diagnosed.

About 70 percent of the time, the use of a logic probe or multimeter is good enough to determine if a computer sensor or other electronic component is working. It's the exceptions that really cause diagnostic headaches. Most home mechanics won't have the need, or can justify the expense, of owning a lab scope. Professional technicians work on a wider variety of vehicles with more computer-related issues; a lab scope can save them hours of time when trying to solve stubborn electrical problems. The price of digital lab scopes has come down in recent years, but they're still expensive—ranging from $800 to well over $3,000. Search the Internet under "automotive lab scope" and you'll get an idea of what's available.

There is a new, interesting alternative to a digital lab scope if you already own a personal computer (PC), laptop, or personal digital assistant (PDA)—a scope interface (a box with leads coming out of it and a universal serial bus, or USB, port). It plugs into a PC or PDA and its included software lets the user view, record, and store waveforms. OTC makes a Palm Scope interface (part number 3961) for use with a Handspring Visor PDA. Other software and interface devices typically run about $300 and can be found on the Internet.

Another alternative is to use an analog lab scope. Unfortunately, these tools are not designed for automotive use and do have limitations. Analog scopes operate in real time so they cannot record waveforms, consequently making it difficult to find intermittent electrical glitches. Analog scopes can be purchased new for as little as $350, and even less, if used.

SOLDERING TOOLS

Having the right tools to repair electrical problems is as important as having the right electronic testing tools to find problems in the first place. Basic soldering tools are essential for repairing and splicing wires, adding connectors, and overall general electrical repair. The electric soldering gun

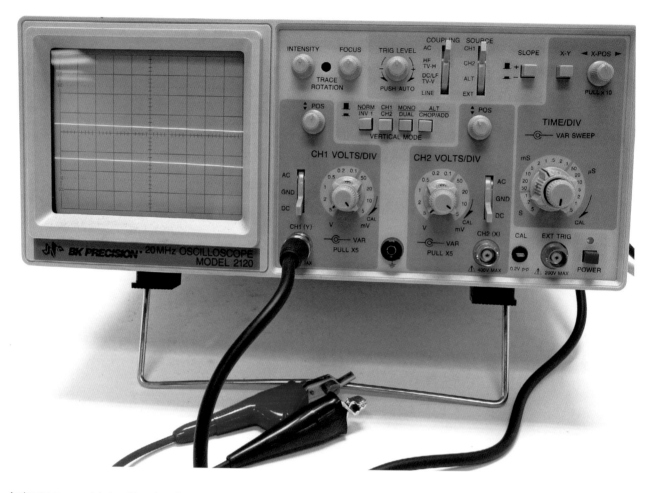

Analog scopes are not designed for automotive use and have limitations. They can only operate in real time and can't record waveforms. They can be purchased new for as little as $350.

Weller's Heavy-Duty Soldering Iron Kit (part number SPG80L) produces 80 watts of power and comes with two tips. A safety indicator light lets you know the iron is on. Courtesy of Weller

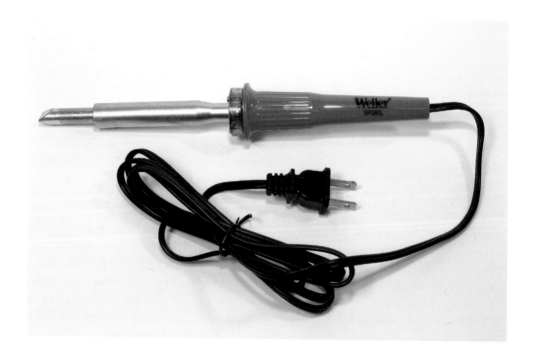

This Weller Dual Heat Range Soldering Gun Kit (model 8200PK) is standard issue for automotive electrical work. Soldering guns heat up quickly and are easy to work with; some even have a work light to illuminate what you're soldering. Courtesy of Weller

Weller's Portasol—a completely portable, butane-powered soldering iron. This tool is convenient and invaluable when working on trailer wiring, under the dash, at the racetrack, or on those dreaded roadside repairs. Heat is adjustable and ignition is via a piezoelectric system—just flick the ON switch. The tool can be charged with butane using the same canisters that fill cigarette lighters. One fill-up lasts several hours—more than enough time to solder entire wiring harnesses. Courtesy of Weller

has been around for years—owning a quality one will make all electrical work go smoothly. Weller, a division of Cooper Tools, has manufactured a variety of soldering tools for many years and is considered the standard in both automotive and electronics industries.

The Weller 8200PK multipurpose soldering gun kit features a dual heat range (100/140 watts) element. By pulling the trigger to the first position, 100 watts of power and heat are produced at the gun tip. This setting works well for smaller wires (up to 14 gauge). When working with larger wires (10 and 12 gauge), the second trigger position is capable of producing 140 watts. This range works best for most automotive electrical work. If you need to work with heavier wire or terminals, Weller D550PK develops 200/260 watts of power. Both models have built-in lights—great when working in dark spaces under a dash or hood. However, since most available spaces on a vehicle are not large enough to accommodate a soldering gun, a pencil size soldering iron is also available to get into those cramped quarters much easier. In addition to standard electric soldering guns, Weller also makes butane-powered soldering irons for the ultimate in portability.

SECTION III
ELECTRICAL SYSTEMS

CHAPTER 4
STORAGE BATTERY

Like most people, you probably don't spend a lot of time worrying about the battery in your vehicle, until the engine won't start; then the battery gets 100 percent of your attention. Fortunately, an engine can usually be jump-started by another vehicle, but the bigger problem is whether it will start again once you've reached your destination. Charging the battery may be only a temporary fix. In addition to not starting an engine, a weak or old battery can also cause drivability issues with EFI vehicles, as well as problems with the charging system. In fact, Chapter 5, Charging and Starting Systems, stresses that the battery should always be fully charged and tested before diagnosing potential charging-related problems in a vehicle. Understanding how a battery works and how to test it will help when you attempt to solve electrical-related issues in your car or truck.

This Optima Red Top battery has 800 cold cranking amps (CCA) and 110 minutes of reserve capacity. The battery is leak-proof and can be mounted anywhere inside a vehicle and in almost any position. Courtesy of Optima Batteries and Summit Racing Equipment

Batteries have three basic functions: (1) to provide electrical power to start the engine, (2) to supply additional current when the charging system can't keep up with electrical demand, and (3) to act as a voltage stabilizer for the vehicle's charging system.

A battery's primary job is to start the engine, and in this case, size does matter. Engine displacement is the key factor in determining amperage requirements, which varies according to the type of engine installed for engine starting and related battery capacity: four-cylinder engines use between 50 and 150 amps, six-cylinders and small V-8s need between 75 and 175 amps, while large V-8 engines can use as much as 275 amps for starter cranking. A battery's cold-cranking amperage rating is directly proportional to the vehicle's engine size since every engine requires a minimum level of amperage to start, which must be met or exceeded by the battery's capacity. Batteries with less capacity are capable of starting a large displacement engine, but not reliably, especially in colder climates. How batteries are rated is covered later in this chapter. Other factors contributing to starter current demand (and related battery size) include: engine/starter cranking ratio, internal starter gear reduction, oil viscosity, ambient temperature, and overall starter circuit resistance.

A battery's second job is to supply current when a charging system is overworked. This usually occurs (though not always) when the engine is idle. If a vehicle's electrical system is creating a high demand (headlights on, heater blowers on high, windshield wipers operating), and engine speed is too low for the alternator/generator to supply enough current, the battery will make up the difference. This can also occur whenever aftermarket electrical accessories have been installed, since high-wattage sound systems, driving lights, camper or trailer hookups, or other accessories can all exceed the charging system's current-producing capacity. When the alternator/generator is at maximum capacity, the battery supplies any additional current required. But only for a short time! If the charging

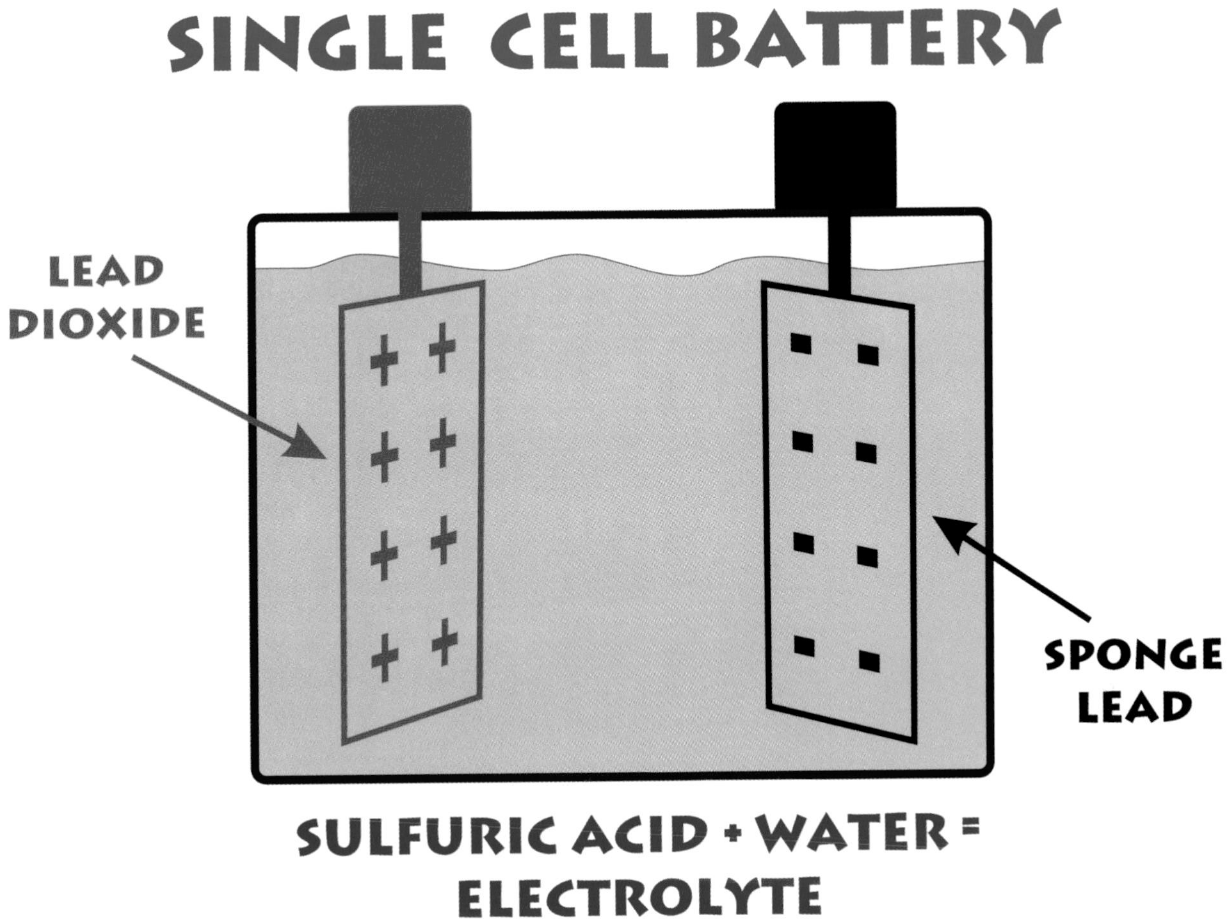

SINGLE CELL BATTERY

LEAD
DIOXIDE

SPONGE
LEAD

SULFURIC ACID + WATER =
ELECTROLYTE

Fig 4-1. *This single cell battery has one positive and one negative plate. The plates swim in an electrolyte solution of sulfuric acid and distilled water.*

system is already unable to keep up with the vehicle's electrical loads, it can't also supply battery charging current for long.

The last function of a battery is to act as a voltage stabilizer for the charging system. The alternator needs something to "push" against to keep from producing excessive voltage. As a result, a battery should never be disconnected on a vehicle equipped with an alternator since the charging output voltage can increase to over 20 volts—enough electrical pressure to take out many (if not all) solid state components like ignition modules, computers, and stereos.

In addition to acting as a voltage stabilizer, a battery provides various electrical system protections. High voltage spikes may be produced when turning on or off certain electrical circuits; these fluctuations in voltage are partially absorbed by the battery, thus protecting solid state components from damage.

CHEMICAL REACTIONS

A battery is basically nothing more than a simple chemistry set that stores electrical energy. It's important to understand a battery does not store electricity; rather, it stores chemical energy necessary to produce electricity. When a battery produces current (to start an engine, for instance), it converts its chemical energy into electricity by a simple change in the form of energy stored. While the process may seem mysterious (since it occurs inside a battery), the following explanation should help take the mystery out of it.

Batteries are made up of a series of separate compartments consisting of pairs of negative and positive lead plates called cells. Each cell has two sets of lead plates—one made from lead dioxide with a positive charge and the other from metallic sponge lead with a negative charge. The plates are stacked alternately (negative, positive) and immersed in an electrolytic solution of sulfuric acid and distilled water. The active material (lead dioxide in the positive plate and

59

SIX CELL 12 VOLT BATTERY

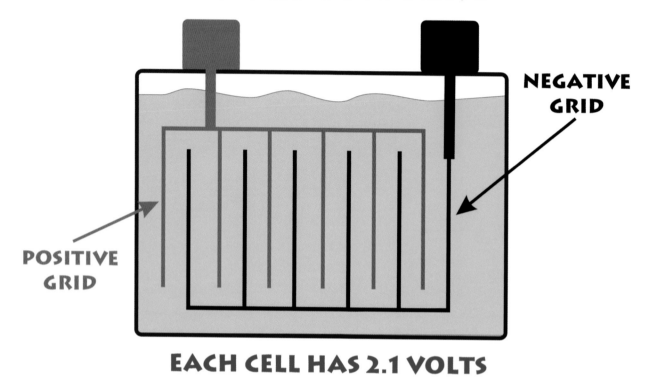

NEGATIVE GRID

POSITIVE GRID

EACH CELL HAS 2.1 VOLTS

Fig 4-2. *Six cells, producing 2.1 volts each, are connected in series to create a 12-volt battery. Six-volt batteries have only three cells.*

metallic sponge lead in the negative plate) produces electricity when immersed in electrolytic solution.

Each cell in a battery produces 2.1 volts DC, and a 6-volt battery has three cells. The overall surface area of the lead plates in the cells is supported by a cast lead framework called a grid, which not only holds the plates together, but also facilitates the flow of electrons between negative and positive plates during charging and discharging. The amount of electrical energy a battery is capable of producing is dependent upon the surface area of the plates in the cells. However, while size does matter to some extent, plate design and specific chemistry are more determinative of the amount of electrical energy a battery can produce. This is why a physically small battery can have an amperage rating higher than that of a battery of larger dimensions.

DISCHARGING

One of two chemical processes is always occurring inside a battery at any given time—either charging or

discharging. The electrolyte solution contains charged atomic particles called ions, which are made up of sulfate and hydrogen. The sulfate ions are negatively charged, while the hydrogen ions have a positive charge. When you place a load across a battery (starter motor, headlight, or horn) the sulfate ions travel to the negative plates and give up their negative charge, causing the battery to discharge. This excess electron flow out of the negative side of the battery, through the electrical device, and back to the positive side of the battery is what creates DC current. Once the electrons arrive back at the positive battery terminal, they travel back into the cells and reattach to the positive plates. This process of continuous discharge continues until the battery is dead and there is no chemical energy left. This flow of electrons from negative to positive battery posts contradicts popular notions about electricity moving from the positive post of a battery through an electrical device and back to the negative post (conventional electron theory—see Chapter 1).

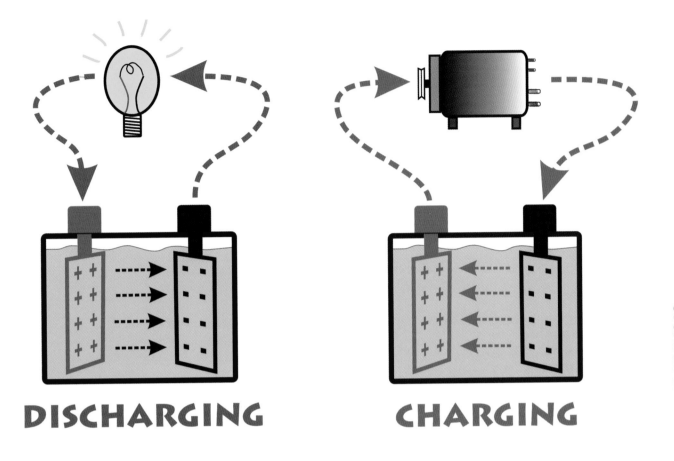

DISCHARGING　　**CHARGING**

Fig 4-3. *The battery on the left is discharging as its positive ions travel from the positive plate to the negative plate. The process is reversed (on the right) when the battery is being charged.*

When a battery discharges, the ratio of sulfuric acid to water in the electrolytic solution changes to mostly water. At the same time, battery acid produces a chemical byproduct called lead sulfate that coats the plates, thereby reducing the surface area over which chemical reactions can occur. As the surface area of the battery plate shrinks, current-producing capacity lessens as well. This is why a battery goes dead after a load device is left on for prolonged periods of time.

Speaking of discharging, a common misconception about battery storage is that if it is left on a concrete floor, it will eventually discharge. This was true when battery cases were made of hard rubber (about 35 years ago); the moisture from concrete caused the battery to discharge through the case. However, modern battery cases are made of acrylonitrile-butadiene-styrene (ABS) plastic, which can be stored on concrete indefinitely without any concern for discharge potential.

ELECTROLYTE SOLUTION

As previously mentioned, a battery's negative and positive plates swim in an electrolytic solution of sulfuric acid and water, an ingredient necessary for electrochemical processes to occur inside a battery. Whenever the ratio of sulfuric acid to water is measured, the resulting value is expressed as specific gravity (SG). The specific gravity for pure water is 1.000, while sulfuric acid has a specific gravity of 1.835. Combined, their specific gravity is 1.265 (indicating a fully charged battery at 80 degrees Fahrenheit). However, as a battery discharges, the ratio of acid to water changes; there is less sulfuric acid and more water, so the specific gravity of the electrolyte solution lowers. The process is reversed when the battery is charged. The specific gravity gets higher as the ratio of acid to water changes back to mostly acid.

Temperature also has an effect on a battery's ability to produce current—the lower the temperature, the lower the specific gravity, and the lower the current-producing

This battery charging guide uses a battery's reserve capacity to determine how long a fully discharged battery should be charged. Both high and low charge rates are listed. Courtesy of Battery Council International

Battery Charging Guide
Charge rates and times are for fully discharged conditions

Rated Battery Reserve (in minutes)	Slow Charge	Fast Charge
80 minutes or less	15 hours @ 3 amps 1.5 hours @ 30 amps	2.5 hours @ 20 amps
80 to 125 minutes	21 hours @ 4 amps 1.5 hours @ 50 amps	3.75 hours @ 20 amps
125 to 170 minutes	22 hours @ 5 amps 2 hours @ 50 amps	5 hours @ 20 amps
170 to 250 minutes	23 hours @ 6 amps 3 hours @ 50 amps	7.5 hours @ 20 amps
Greater than 250 minutes	24 hours @ 10 amps 4 hours @ 60 amps	6 hours @ 40 amps

potential. There are specialty batteries that use different specific gravity specifications based upon the range of temperatures in which the battery is designed to operate. Batteries used in tropical climates (where ambient temperatures never freeze water) have lower specific gravity values, usually of 1.210. Batteries used in extremely cold climates use higher specific gravity values of up to 1.300. Higher specific gravity values are typically not found in batteries intended for use in North America or most of Europe because battery life is significantly decreased by a higher specific gravity rating. A battery with a specific gravity of 1.265 typically will not freeze within the continental United States because the electrolyte solution can be subjected to –75 degrees Fahrenheit before it will solidify. However, batteries should always be kept charged in cold climates to prevent damage. As a battery becomes discharged, its specific gravity is lowered, and so is the freezing point of its electrolytic solution is raised. A dead battery with an specific gravity of 1.155 has a freezing point of about 5 degrees Fahrenheit, at which point it could freeze solid and damage the battery.

CHARGING

Charging a battery reverses the chemical reactions that occurred during discharge. Basically, the sulfate and hydrogen ions switch places. The electrical energy used to charge a battery is converted back into chemical energy and stored inside the battery. Battery chargers, including alternators and generators, produce a higher voltage (higher electrical pressure) than the open circuit voltage of the battery they are charging. This electrical pressure is required to push the current back into the battery (overcoming its open circuit voltage) in order to charge it. The charging device (alternator, generator, or battery charger) produces excess electrons at the negative battery plates, where positive hydrogen ions are then attracted to them. The hydrogen ions combine to form sulfuric acid and lead, which ultimately reduces the amount of water in the electrolytic solution and increases the battery's specific gravity.

Applying a charging current to a battery without overheating it is called the natural absorption rate. When charging amperage exceeds the level of the natural absorption rate, the battery may overheat, causing the electrolyte solution to bubble and create flammable hydrogen gas. When combined with oxygen from the air, hydrogen gas is extremely explosive and can easily be ignited by a spark. Therefore, you must always remember to turn the power off before connecting or disconnecting a battery charger in order to prevent a spark at the battery terminals! Some newer battery chargers have a no-spark feature, even when disconnected under power. A battery that has been rapidly discharging (due to cranking an engine over until the battery died) may also produce excessive hydrogen gas.

A spark can also result when connecting jumper cables to a dead battery, which can also cause an explosion. Figure 4-4 illustrates the correct method for connecting jumper cables to a dead battery when attempting to jump-start your vehicle with another vehicle. Always follow these four steps when jump-starting a vehicle: (1) connect the red jumper cable to the dead battery's positive terminal, (2) attach the other end of this same cable to the charged battery's positive terminal, (3) connect the black cable to the good battery's negative terminal, and (4) connect the other end of the negative black jumper cable to the dead vehicle's engine block or frame. The last connection is the one that sparks. This method will keep all potential sparks away from both batteries.

Automotive batteries can be charged at a high rate (over 60 amps) as long as the battery case temperature does not exceed 125 degrees Fahrenheit. To determine if this temperature is being exceeded, place your hand against the battery case; if you can't leave it there for more than a few seconds, the temperature is too high and the charging rate should be lowered. Also, if the battery's cells are producing excessive gas (evidenced by vigorous bubbling of the electrolyte solution), the charging rate should be reduced. For these reasons, high-rate or boost-fast battery chargers should not be left unattended for long periods of time. Automatic, smart battery chargers are able to sense battery voltage and shut themselves off when the battery reaches full charge. Many of these smart chargers also change to trickle charge mode in order to maintain a battery's state-of-charge over long periods of time. A dedicated trickle charger will not charge a dead automotive battery since most of these chargers only have a charging rate of 1 amp; however, they are ideal for use on batteries stored during winter or used in boats, lawn equipment, and motorcycles.

BATTERY RATINGS

Since a battery's basic job is to power the starter motor while maintaining sufficient voltage to also run the ignition and fuel systems, it is rated in cold cranking amps (CCA). CCA represents the discharge load in amps that a new, fully charged battery at 0 degrees Fahrenheit can continuously deliver for 30 seconds while maintaining 7.2 volts. CCA battery ratings for cars and light trucks are generally in the range of 300 to 800. More expensive batteries have greater CCA capacity; some are as high as 1,000.

Another type of battery rating measures reserve capacity (RC)—a battery's ability to power a minimum electrical load if the vehicle's charging systems quit working. Generally, this load is defined as enough current to operate the vehicle with the headlights on low beam and the windshield wipers functioning. RC is rated in the numbers of minutes a fully charged battery can discharge 25 amps and maintain 10.5 or greater voltage at 80 degrees Fahrenheit. Battery RCs range from 45 to over 250 minutes and correspond with CCA ratings. The old adage, "you get what you pay for," applies to the purchase of automotive batteries, as higher CCA-rated batteries are warranted for more years than those with less CCA and have greater RC as well. RC may also be used as a guide for charging rates (see page 62).

TYPES OF BATTERIES

There are three types of batteries available for cars and light trucks—conventional, absorbent glass mat (AGM), and gel. Because a battery can potentially leave you stranded,

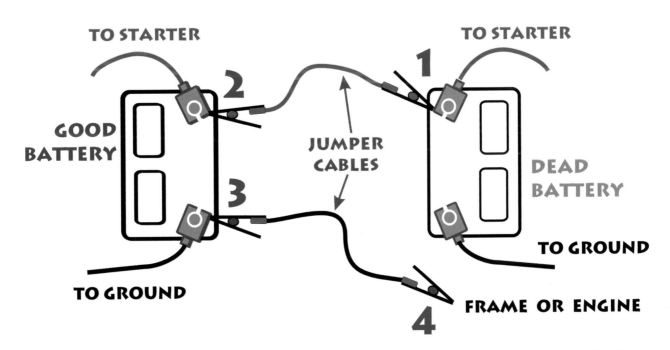

Fig 4-4. *For safety when attempting to jump start your vehicle, always connect the last jumper cable (step 4) to the engine block, or frame of the vehicle, with the dead battery. This will prevent a spark from occurring at the negative terminal of the dead battery.*

This Moroso power charger will charge 12- and 16-volt batteries at up to 30 amps. It can be set for AGM, conventional, and deep-cycle types of batteries. The charger will automatically shut down if connected to a battery backward. Courtesy of Moroso and Summitt Racing

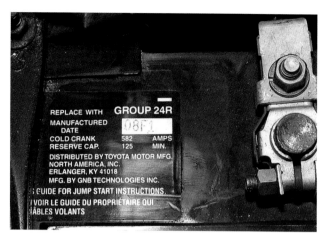

The CCA rating on this Toyota battery is listed as 582. Reserve capacity is 125 minutes. This battery is adequate for a small, four-cylinder engine. Courtesy of Younger Toyota

you should always buy only high-quality brands. Rebuilt or bargain basement batteries may start your vehicle for a while, but they are sure to leave you stranded sooner rather than later.

Conventional Batteries

Sometimes referred to as flooded batteries, these are the most basic design. Batteries made in earlier times used lead-antimony as a material for the plates. Today, conventional batteries are made with lead calcium, adding much improved self-discharge properties. When the battery discharges and charges, some of the water in the electrolyte solution evaporates, causing the level of cells to become low. These batteries have filler caps on top, allowing you to check each cell's electrolyte level. The individual cells periodically need to be topped off with distilled water due to inevitable water loss during the charging/discharging process. When charging a conventional battery, always remove the caps to let the hydrogen gas escape.

Absorbent Glass Mat (AGM) Batteries

Unlike conventional batteries, AGM batteries are designed so there is no free, unabsorbed electrolyte solution to spill or evaporate, nor do they need to have water added. Hence, these batteries are commonly described as maintenance-free. The electrolytic soup is fully absorbed and permanently held in capillary attraction by either glass-fiber or wool-mat separators between the lead plates or grids. This design gives antivibrational support to the battery plates, keeping them from short circuiting between negative and positive grids. In addition, AGM batteries can

be physically smaller, yet provide the same or more cranking power than conventional designs.

Gel Batteries

Gel batteries contain electrolyte solution in gel form but have conventional construction, with paper or polyethylene insulators separating the plates. The electrolytic solution is a mixture of sulfuric acid, water, and silica and the result is a toothpaste-like consistency. Until recently, gel-electrolyte construction was limited to what was known as deep-cycle batteries. Deep-cycle batteries can be used until dead and recharged with little loss in overall capacity. Conventional and AGM batteries have shorter life spans if allowed to completely discharge. Gel batteries are primarily used in mobility equipment (wheelchairs), golf carts, recreational vehicles (RVs), marine applications, and automotive racing. The bottom line on gel batteries is that they have higher internal electrical resistance than AGM batteries and don't offer as high a rate of performance. In other words, what works well on your golf cart or race car won't necessarily start a medium- to large-displacement engine in your vehicle over the long haul, especially in cold weather.

BATTERY TESTING—STATE OF CHARGE

There are several methods available for testing a battery, ranging from expensive, professional-grade test equipment to a multimeter and the engine's starter motor. All of the tests, with one exception, require the battery to be fully charged. If the battery is a conventional type, the filler caps can be removed and a hydrometer test can be used to determine state-of-charge by measuring specific gravity.

Some auto parts stores still carry hydrometers, but since they are gradually being phased out, they can be hard to find. Most batteries used today are maintenance-free types, which have no filler caps for hydrometer testing. However, instead of measuring specific gravity, an open circuit voltage test can determine state-of-charge on both conventional and maintenance-free batteries.

A battery has to "rest" for at least 10 minutes with no load before performing an open circuit voltage test. Make sure all electrical loads are turned off before testing. Connect a digital multimeter directly to the battery and measure voltage. (See the chart on page 66 to determine state of charge.) If the state of charge is less than 75 percent, the battery must be charged before further testing. After charging, if voltage does not increase to 12.6 or higher, you need to replace the battery. If the state of charge is more than 75 percent, the surface charge must be removed before

further testing. (All of the stated battery-testing voltage values are for 12-volt batteries; when testing 6-volt batteries, divide voltage specifications in half.)

DYNAMIC BATTERY TESTING

If the state of charge is more than 75 percent, operating the starter motor will remove the surface charge. Disable the ignition and fuel injection systems (if equipped); this can be done on late model vehicles by finding the ignition or EFI fuses and removing them. On carbureted vehicles, disconnect the coil wire at the distributor cap, and ground the coil wire by using a jumper wire. Don't just unplug the coil wire and let it dangle on an electronic ignition system; the ignition module can get fried if the spark has no ground return. Crank the engine for 15 seconds, and then let the battery rest for 15 seconds. This will remove the surface charge. With a voltmeter connected to the battery, crank

After cranking the engine for 10 seconds and with the starter still cranking, this battery is showing 10.46 volts, which indicates that it is in good condition.

Measuring open circuit voltage is an accurate method for determining a battery's state of charge. Make sure all electrical loads are off when performing the test. Courtesy of Battery Council International

Open Circuit Voltage and State of Charge

State of Charge	Voltage (12v)	Voltage (6v)
100%	12.65 +	6.32
75%	12.45	6.21
50%	12.24	6.12
25%	12.06	6.02
Discharged	11.89	5.93

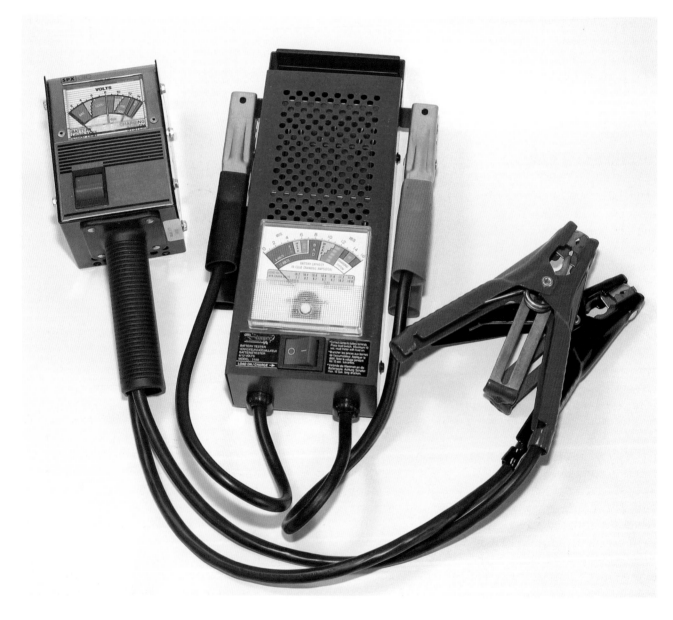

Hand-held battery testers are easy tools to use to perform a battery load. The heavy-duty OTC battery tester on the left (part number 3181) will load test a battery at 130 amps and has clamps that work on both top- and side-post battery terminals. . OTC's Stinger battery tester (part number 3180, at right) draws a 100-amp during load testing. Courtesy of SPX/OTC Service Solutions

the engine for an additional 15 seconds. Watch the voltmeter reading just before you stop cranking the starter; a good battery should have a minimum loaded voltage of 9.6 volts at 70 degrees Fahrenheit. If the testing is done in a cold climate (an ambient temperature of about 40 degrees Fahrenheit), the minimum voltage should read 9.3 volts. If the loaded voltage is less than minimum, the battery is weak or tired and on its way to becoming a starting problem. If the battery is bad, loaded voltage will drop way off (less than 7 volts) in the first few seconds of engine cranking.

HAND-HELD BATTERY TESTERS

Hand-held battery testers are becoming more widely available and can now be found in most auto parts stores. An OTC Stinger battery tester (part number 3180) is an example of a quality hand-held tester. Instead of using the starter motor, a hand-held tester is used to load the battery. After connecting the tester to the battery terminals, a spring-loaded switch on the tester triggers an internal relay. The

Depending on CCA, loaded voltage in the range of 9.4 to 11.2 indicates a good 12-volt battery. The OTC Stinger battery tester can also test 6-volt batteries using a different scale on the meter's face.
Courtesy of SPX/OTC Service Solutions

relay connects the heavy tester cables to the battery and places a 100-amp load across the terminals. The tester meter gives a reading of the battery condition.

When using a Stinger battery tester, connect the red and black cable clamps to the positive and negative battery terminals. The tester's meter will indicate state of charge (see page 66). The battery's CCA can usually be determined by reading the information label on the battery. If the rating is not printed on the battery, use the following as a guide to estimate CCA: small four-cylinder engine—300 CCA; medium V-6—400 CCA; and large V-8—600 CCA. While watching the meter, depress the load switch on the tester for 10 seconds. At the end of 10 seconds, read the meter with the switch still depressed. Use the scale corresponding with the CCA of the battery being tested. If the needle is

in the green area, the battery is good. A steady needle reading in the yellow area means the battery may need to be charged or it could cause starting problems in the future, especially in cold weather. If the needle moves noticeably down the scale, the battery is bad and needs to be replaced. To compensate for testing in cold ambient temperatures, the CCA scale consulted should be lower. For example, at 50 degrees Fahrenheit, the CCA scale used should be 100 CCA less than the battery's rating.

The previous battery testing methods required a fully charged battery before testing. However, a professional-grade battery tester is capable of testing a battery even if it's dead. These digital testers measure a battery's internal resistance regardless of state-of-charge. Internal resistance is an indication of a battery's ability to deliver current. The

This digital battery tester can even test a dead battery. The tester uses battery capacitance to determine if a battery is good. These units are expensive and used mostly by professional technicians.

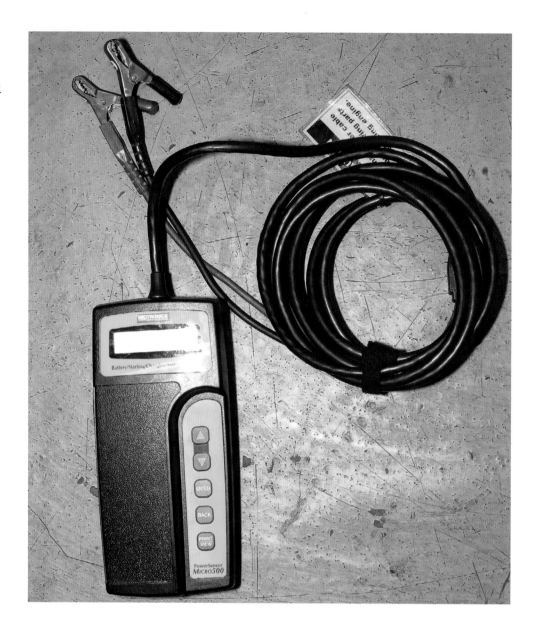

more capacity a battery has to produce amperage, the lower its internal resistance. A digital capacitance battery tester uses single-load dynamic resistance technology to calculate battery performance. These testers use a modified DC load test to apply a small, momentary load to the battery while measuring instantaneous voltage drop across all cells. The load is then removed and voltage across the cells is measured again after a recovery period. These analog measurements are converted into digital information; the tester calculates

the dynamic internal resistance in order to evaluate overall condition. The entire process takes about two seconds and current drain on the battery is minimized. These testers provide information on open circuit voltage, state of charge, and battery health and condition. They can also test a partially charged or fully discharged battery whether on or off the vehicle. The only drawback is the price—these testers are not cheap! Thus, the cost is usually only justifiable for professional technicians.

CHAPTER 5
CHARGING AND STARTING SYSTEMS

CHARGING SYSTEMS

The previous chapter focused on automotive storage batteries and battery chargers in general. In this chapter, the focus shifts to an onboard battery charger, commonly known as an alternator (or a DC generator in older cars). The charging system is the heart and soul of the automotive electrical system. Without a reliable charging system, anything in a vehicle requiring electricity will not work for long.

In the early part of the 1900s, before batteries were installed in cars or motorcycles, only ignition systems required electricity to operate. They used a magnito—a small generator capable of producing just enough energy to operate only an ignition coil. Once electric lights and starters were introduced, so were storage batteries; however, there had to be some way to keep them charged. DC generators were used to charge batteries until the early 1960s when Chrysler produced an AC generator (better known as an alternator). By the mid-1960s, most generators had been phased out of production. The last automobile manufacturer to use a generator was Volkswagen in 1973.

Alternators represented new technology in the 1960s—only possible because of developing solid state electronics such as transistors and diodes. Since an alternator must convert AC voltage output into DC volts, the diode proved to be an economical means of executing this process. Early alternators used mechanical voltage regulators to control outputs, but these were soon replaced with solid-state devices. In the early 1980s, Chrysler incorporated a voltage regulator for the charging system into its onboard computer. Charging system malfunctions were then identified via onboard computer self-diagnostics, which set trouble codes to assist technicians. Today, most automobile manufacturers use an engine management computer to control, monitor, and diagnose the charging system.

The charging system has only two purposes: to charge the vehicle's battery and to power all the electrical components in the vehicle once the engine starts. The charging system must have enough capacity to meet all the electrical demands of the vehicle. Both DC generators and alternators are rated for the amount of current they can produce. If a car's total electrical power requirement is 55 amps (including charging the battery), the charging system

This powder-coated Magnum alternator is available in an 80-amp, single-wire version specifically designed for 1972–1979 GM vehicles; there is a similar 65-amp version for 1961–1986 Ford vehicles. This alternator exceeds original equipment manufacturer (OEM) specifications and comes in various colors. Courtesy of Magnum and Summit Racing Equipment

must produce at least that much current to keep the battery from going dead. Manufacturers don't want charging systems operating at maximum power output for long periods of time, so they install larger capacity alternators or generators. For a vehicle with total electrical power requirement of 55 amps, an alternator capable of producing 60 amps would be used. If a vehicle's electrical system requires more current than the charging system is capable of producing, the storage battery will make up the difference until it goes dead. Naturally, this process is accelerated if the charging system has a problem. Not only will electrical components steal power from the battery to keep operating, but a malfunctioning charging system won't recharge the battery either.

DC GENERATORS

Generators produce electricity by means of a magnet and coil of wire. All magnets have north and south poles, which

DC generators are limited in current-producing capacity because all amperage must pass through the generator brushes. Most generators are only capable of producing about 25 to 40 amps—not enough for modern vehicles. Courtesy of Auto Electric, Inc.

create an invisible field around the magnet. When a coil of wire is rotated between two magnets, voltage is induced into the wire; similarly, the magnets can be moved and the coil of wire kept stationary with the same results. This process of rotating either wire coils or magnets is called magnetic induction. In a simple generator, the magnets are inverted and separated by an air gap, so the north pole of one faces the south pole of the other. The magnetic forces from one magnet bridge the air gap and extend to the other magnet. When a coil of wire is placed between the two stationary magnets and rotated, the magnetic lines of force are intermittently cut. Each time the magnetic lines of force are interrupted, voltage and current are induced into the coil of wire. When the coil of wire is rotated 180 degrees, the lines of force are cut in the opposite direction, causing induced voltage and current to reverse direction. This back and forth reversal of current and voltage is known as alternating current (AC). Both DC generators and AC generators (alternators) produce AC voltage and current. However, automotive electrical systems only use DC output and voltage. AC output from either a generator or

an alternator must be converted into DC, but alternators and generators accomplish the conversion of AC to DC in different ways.

A DC generator produces electrical current when coils of wire rotate past electromagnetic stationary field coils. Multiple coils of wire are wrapped around a laminated iron core called an armature, which rotates via a belt powered by the engine's crankshaft. The circular ends, or loops, of each coil of wire are isolated and connected to copper commutator bars located at the end of the armature and positioned in a circle around the armature shaft. The stationary field coils, which act like magnets inside a generator, are actually electromagnets in the form of coils of wire wrapped around iron-pole shoes attached to the inside of the generator's case. The field coils are energized by the battery when the ignition key is turned on; once the engine is running, the generator self-powers the field coils. As the armature turns, AC voltage is produced and then converted into DC voltage via two carbon-based generator brushes that act like mechanical switches as they connect with the rotating commutator bars. The generator brushes are mounted 180 degrees apart, allowing contact with only one armature wire loop at a time. As each wire loop rotates past the brushes, a small voltage passes into the brush and then into the car's electrical system. Because the commutator bars rotate rapidly across the brushes, voltage output is constant. Since the commutator bars have to carry the generator's output, the amount of current that both 12- and 6-volt DC generators can produce is limited. Heavy-duty generators can only produce about 50 amps, while a typical DC generator produces between 25 to 45 amps.

DC GENERATOR VOLTAGE REGULATORS

As a DC generator rotates, voltage output and current increase. If left unchecked, it would do so until the generator self-destructs. Consequently, DC generators are equipped with an external mechanical voltage regulator that controls both current and voltage. When output voltage becomes too high, the voltage regulator switches off the field coils in the generator. Inside the voltage regulator is a calibrated electromagnet that moves a set of contact points (similar to those used in older ignition systems) open and closed; these turn the field coils on and off at a rate of about 200 times per second. With no current going to the field coils, no magnetic lines of force are cut inside the generator, and it stops producing current. As voltage in the electrical system drops, the regulator switches the field coils back on. By turning the field coils inside the generator on and off, a voltage regulator can set voltage at a constant rate, even though the vehicle's electrical loads may vary.

If the battery is discharged or there are excessive electrical loads placed on the charging system, the voltage regulator will keep the generator's field coils energized all the time in an effort to keep up with the loads. In addition to controlling voltage, the voltage regulator limits the amount of current the generator produces. The current regulator circuit inside the voltage regulator works in a manner similar to the voltage regulator—as amperage flow increases from the generator, an electromagnetic coil pulls open a second set of contact points and turns off the field coils, limiting the generator current output.

A voltage regulator has an additional component known as a cut-out relay, which prevents the battery from discharging back into the generator's field coils when the key is turned to OFF or the engine is idling. Without the cut-out relay, the generator would always be on and the battery would go dead, even with the engine not running. Older voltage regulators could be adjusted by bending the metal tabs that held the springs for the contact points. The tabs were eliminated and later replaced with nonadjustable regulators, meaning that when the voltage was too high or low, the only option was replacement. While modern solid state electronic voltage regulators use diodes and transistors instead of springs, coils, and contact points, they are still also nonadjustable.

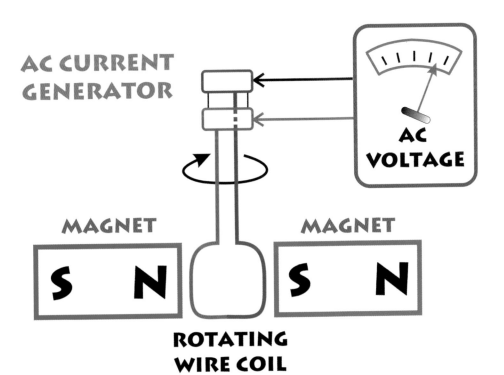

Fig 5-1. *Both generators and alternators work in a similar manner. A simple coil of wire is rotated between two magnets. As the invisible lines of magnetic force are cut, voltage and current are induced into the coil of wire.*

This diode trio is used in an alternator to rectify AC into DC. Because the diodes get hot during operation, an external fan is used to create airflow inside the alternator. Courtesy of Auto Electric, Inc.

ALTERNATORS

The development of a solid state device called a diode led to the wide use of alternators in the early 1960s. A diode is an electrical one-way valve that allows current to pass in one direction, but not the other. The diodes inside an alternator convert AC and voltage into DC by means of a process called rectification.

An alternator produces AC in the same manner as a DC generator by the use of electromagnets and coils of wire. However, a major difference between DC generators and alternators is the relative positions of the magnets and coils of wire. An alternator has a rotating magnet and stationary coils, while a DC generator uses rotating coils with stationary magnets or field coils. Also, the rotor inside an alternator takes the place of an armature. The rotor consists of an iron core mounted onto a shaft with a coil of wire wrapped around it, with the coil enclosed between two pole pieces that are also made of iron. The two ends of the coil connect to copper slip rings at one end of the rotor shaft. These slip rings are in contact with two carbon brushes, one of which is grounded while the other is connected through the alternator's field terminal to the voltage regulator. Unlike the segmented commutator bars in a DC generator, slip rings on

This Summit alternator produces 140 amps. A heavy-duty, internal voltage regulator and oversized, American-made bearings are utilized to accommodate the high current output. Courtesy of Summit Racing Equipment

an alternator are continuous. The amount of current passing through the slip rings is far lower than that which passes through the brushes in a DC generator. Unlike generator brushes, alternator brushes usually last the life of the alternator; in fact, other internal components—like bearings and diodes—typically fail before the brushes or slip rings.

An alternator uses three coils of wire that are incorporated into the windings on a stator. The stator coils are fitted

With the rotor removed, the alternator brushes can be seen. They are made from carbon and are spring-loaded so they will make contact with the rotating slip rings when the alternator operates.

The alternator's stator is made up of three separate coils. The rotor spins inside the stator, inducing voltage and current into the stator windings (or coils).

An alternator's rotor consists of a coil of wire and two iron pole pieces. The slip rings that transfer power to the rotor coil are on the right.

THREE AC SINE WAVES

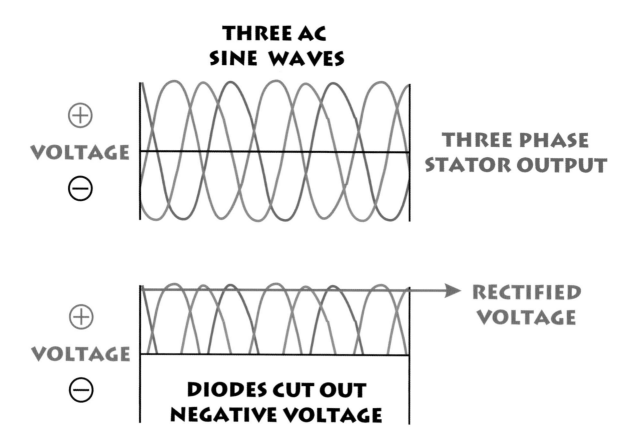

$\oplus$

VOLTAGE

$\ominus$

THREE PHASE STATOR OUTPUT

$\oplus$

VOLTAGE

$\ominus$

RECTIFIED VOLTAGE

DIODES CUT OUT NEGATIVE VOLTAGE

Fig 5-2. The graph at the top represents the voltage output from the three stator coils. Voltage for each coil (a phase) alternates between positive and negative. The lower diagram shows the effect of the presence of negative voltage blocking diodes in the stator circuit. Only the negative part of the stator coil's voltage output is blocked and the resulting voltage is changed into DC voltage.

Fig 5-3. This schematic drawing illustrates how each stator coil's output voltage connects between a pair of diodes. Only positive voltage is allowed to reach the battery as it passes through the three positive (red) diodes.

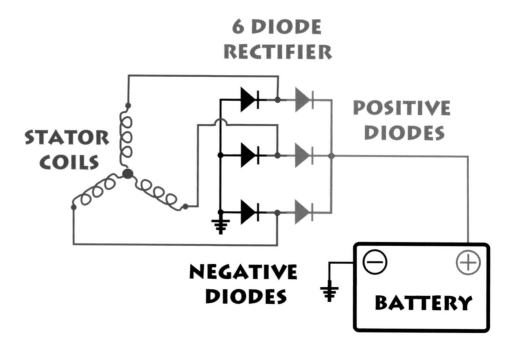

6 DIODE RECTIFIER

POSITIVE DIODES

STATOR COILS

NEGATIVE DIODES

BATTERY

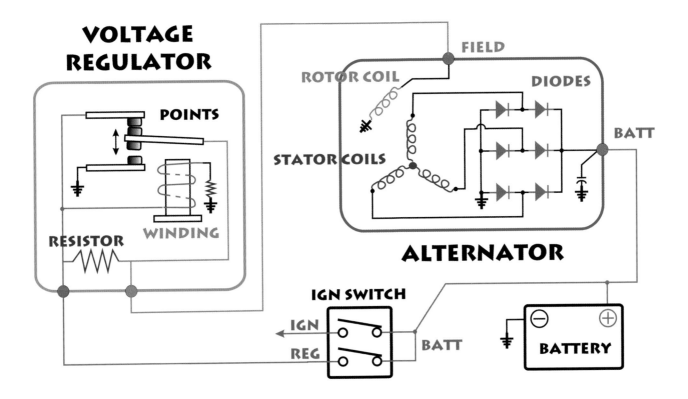

Fig 5-4. *This early alternator uses a mechanical voltage regulator to control the rotor coil. By turning the coil on and off, the alternator's voltage and current output are regulated at a pre-set level.*

inside the alternator and surround the rotor. As the rotor turns, its magnetic field induces voltage into the stator coils. Three coils make up the stator, each producing its own voltage. The stator output is AC voltage that swings up and down between negative and positive in repeating cycles.

Diodes are solid state devices containing no moving parts. They are used to block negative voltage produced in the stator coils. One end of each stator coil is connected to both a positive and a negative diode and the opposite ends of the coils are connected together to form a Y arrangement. Not all alternators are designed this way, but most are. No matter the design, though, they all function in the same manner by using a diode's one-way current-flow attributes to block the negative voltage output from each coil so only positive voltage reaches the vehicle's electrical system.

Alternators have a three-phase output because three coils are used in the stator. These three phases overlap each other, producing (more or less) even DC current and voltage output. This process is also known as AC voltage rectification.

Six diodes are used in an alternator—two for each output phase, or stator coil. Each diode is mounted at the slip ring end of the alternator housing. The three negative diodes are attached to the alternator's frame while the positive diodes are mounted in a heat sink and insulated from ground. Because diodes get hot, an external fan is used to cool things off by creating airflow inside the alternator. (Some alternators use internal fans as well.)

MECHANICAL VOLTAGE REGULATORS FOR ALTERNATORS

Just like DC generators, alternators need voltage regulators to control voltage and current output. Early alternators used external, mechanical voltage regulators to accomplish this. In order for a voltage regulator to control an alternator's output, it must add resistance between the battery and rotor circuit. As the regulator increases resistance to the rotor's circuit, its magnetic field is reduced. As a result, less voltage is induced into the stator coils, and by limiting the voltage reaching the stator windings, current is

Because it has no moving parts, this solid-state voltage regulator is capable of switching on and off about 7,000 times per second. Old mechanical voltage regulators were only able to do this for about 200 times per second and required periodic adjustments and eventually wore out. Courtesy of CARQUEST Auto Parts

Systems incorporating alternators equipped with internal voltage regulators are sometimes referred to as integral charging systems. The alternator and voltage regulator act as a combined single unit, but can oftentimes be serviced separately.

controlled. Figure 5-4 displays how a vibrating contact-point regulator controls an alternator's field coil strength.

A voltage regulator controls the amount of time an alternator's rotor winding is powered by full battery voltage. When the ignition switch is ON, the rotor is directly connected to the vehicle's battery through a stationary upper contact point (see Figure 5-4, page 75). As alternator voltage output increases, the windings inside the voltage regulator produce a magnetic field. At a pre-set voltage level, the regulator pulls a moveable contact down until it touches a lower stationary point. This grounds the rotor coil through the field wire and shuts off power to the rotor. With no power, the rotor's magnetic field also shuts off and alternator output is interrupted or reduced until system voltage drops and the regulator winding releases the moveable contact. Then, the moveable contact point is sprung toward the upper contact, where it receives battery voltage to power the alternator's rotor coil. The contact points inside the regulator vibrate or float between open and closed at a rate of approximately 200 times per second, resulting in steady voltage and current regulation. Figure 5-4 shows a simple regulator through which field current comes directly from the ignition

switch. Other regulator designs use a field relay to connect the rotor to the battery. When this type of alternator starts to turn, a small voltage is induced into the stator windings, which then close the field relay contact points. The field relay provides power to the rotor circuit until the engine is shut off and the alternator stops turning. With no current coming from the alternator, the field relay opens, thereby breaking contact between the battery and rotor windings.

Voltage regulators used in alternators do not require cut-out relays like those found on DC generators because the diodes inside the alternators serve this function by only allowing current to flow in one direction—from alternator to battery. Some alternators come equipped with a diode trio instead, which serves the same function.

TRANSISTORIZED ELECTRONIC VOLTAGE REGULATORS

A transistorized electronic voltage regulator basically performs all the same functions as a mechanical regulator except that a transistor takes the place of coils of wire and moveable contact points. Nearly all external electronic regulators control an alternator's field current by switching

An internal voltage regulator (upper right) is on this GM alternator. Diodes are located on the left; slip ring brushes are at the bottom.

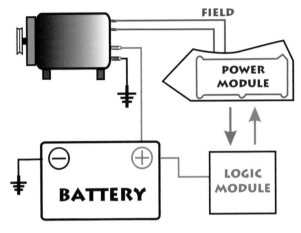

Fig 5-5. Chrysler's early, computer-controlled charging systems used two computers—logic and power modules. The logic module sensed battery voltage and controlled the power module. The power module served as a voltage regulator. These systems were able to compensate for ambient temperatures by increasing charging voltage in cold weather and reducing it under hot conditions.

Fig 5-6. The digital signal from Chrysler's logic module is a series of on and off pulses. This continuous series of signals controls a transistor inside the power module, which in turn controls the alternator's output.

the circuit on and off. Because solid-state electronics (transistors) are used, the switching rate is around 7,000 times per second, compared to a mechanical voltage regulator's contact-points switch-rate of 200 times per second. Obviously, an electronic regulator provides much better control, and all without any moving parts. (Although this is not entirely true, as some early electronic regulators use field relays mounted next to circuit boards. Also some Chrysler voltage regulators sense ambient temperature and adjust the charging system voltage accordingly—higher

voltage for cold weather, lower for hot.)

INTERNAL AND COMPUTER-CONTROLLED VOLTAGE REGULATORS

Some charging systems are designed to allow the incorporation of alternators fitted with internal voltage regulators. These systems are oftentimes referred to as integral charging systems. Automotive manufacturers started using these designs as early as 1973 in order to cut costs and save space. Internal voltage regulators work in basically the same

Chrysler's power module is designed to have all of the air going into the engine pass through its plastic housing. Outside air cools the computer's electronics, including the field coil transistor that controls the alternator from inside the power module. *Courtesy CARQUEST Auto Parts*

manner as transistorized electronic regulators.

Many late model GM alternators equipped with internal regulators have only a single wire leading from the battery to the alternator, making them ideal for use in project vehicles. Chrysler has always been an innovator in charging system design and technology, and it was the first manufacturer to use a vehicle onboard computer to control a charging system. By the mid-1980s, Chrysler's computer-controlled charging system used a logic module and a power module to regulate an alternator field circuit. The logic module senses battery voltage, engine speed, and battery temperature, and then sends a digital signal to the power module, which controls the alternator's field circuit via a large transistor. The power module's transistor sends a digital signal in a continuous series of on and off ground pulses to one side of the alternator's field coil, while the other side receives battery voltage when the key is turned to ON. When engine speed is high, and the electrical load on the charging system is low, the off pulses are longer. However, when electrical accessories are in use (heater blower, windshield wipers), the on pulses are longer, providing more of a ground for the alternator's field coil and increasing charging output. The longer the on cycle, the higher the alternator's output.

CHARGE INDICATORS

Ammeters, voltmeters, and charge indicator lights (idiot lights) are all different instruments that can alert an owner if there is a problem with the charging system. Ammeters,

found in older vehicles, use internal shunts through which all charging system current passes. Since the shunt inside the ammeter has low resistance, it allows most of the current to pass. However, a small amount also passes through the ammeter's windings, producing a magnetic field. The strength of this field moves or deflects a pointer that reads in amps; the greater the amount of current passing through the shunt and ammeter windings, the further the needle or pointer deflects.

Some voltage regulators and alternators are specifically designed for use with an ammeter; however, they are generally not interchangeable with other components in the charging circuit that are not compatible with an ammeter. On some vehicles, the ammeter and alternator are connected in series—if the ammeter stops working, the battery doesn't get charged. Vehicle ammeters usually have a positive and negative scale. When an ammeter's pointer indicates a positive charge, current is flowing from the alternator (or generator) to the battery; but when the needle points toward the negative scale, the charging system isn't producing enough current to charge the battery. Newer cars are almost never equipped with ammeters, since they've typically been replaced by voltmeters or idiot lights.

Unlike ammeters, voltmeters in vehicles have very high resistance; consequently, almost no current passes through them. Voltmeters are connected in parallel to the charging system and measure charging system voltage. A voltmeter uses a small amount of current, which passes through a coil

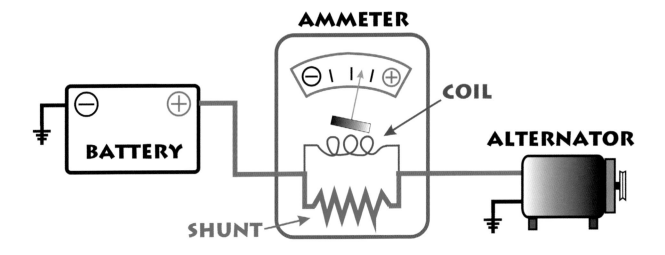

Fig 5-7. *Some older vehicles use ammeters to monitor charging systems. The meter's internal shunt passes charging current from the alternator/generator to the battery. An ammeter uses a pointer to indicate if the system is charging and everything's working, or discharging because the battery is going dead.*

of wire located between two permanent magnets. As the current increases, the coil of wire moves, causing a pointer to move across a scale indicating voltage.

Unfortunately, both in-vehicle voltmeters and ammeters are notoriously inaccurate. Inconsistencies in readings have caused many a headache at automobile dealerships, since customers too often get overly concerned about charging systems not charging at specific levels. Original equipment manufacturers (OEM) solved this problem by switching to idiot lights (charge light indicators) and voltmeters without numerical displays (only HI/LOW instead).

Charge light indicators are usually connected between the ignition switch and voltage regulator (or alternator). Oftentimes, a 500-ohm resistor is connected in parallel with the light as part of the circuit. When an ignition key is first turned to ON, battery voltage is supplied to one side of the charge indicator light, while the other side is connected to the voltage regulator to provide a ground for the charge light when the engine is not running. As soon as the engine starts and the alternator turns, voltage increases on the ground side of the charge light. As voltage builds, the charge light has less and less of a ground, eventually goes dim, and finally turns off. If anything goes wrong in the charging system that causes system voltage to drop, the charge indicator light receives a ground from the voltage regulator and turns on. Unfortunately, in some cars it's common for the charge light to be dim at idle and then shut off when

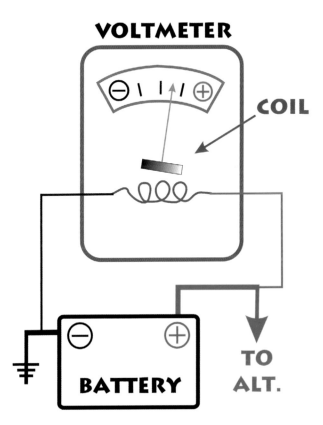

Fig 5-8. *Many modern cars use voltmeters on instrument panels to indicate if charging systems are working. Unlike instrument panel ammeters, a voltmeter is not connected in series, and almost no current passes through it.*

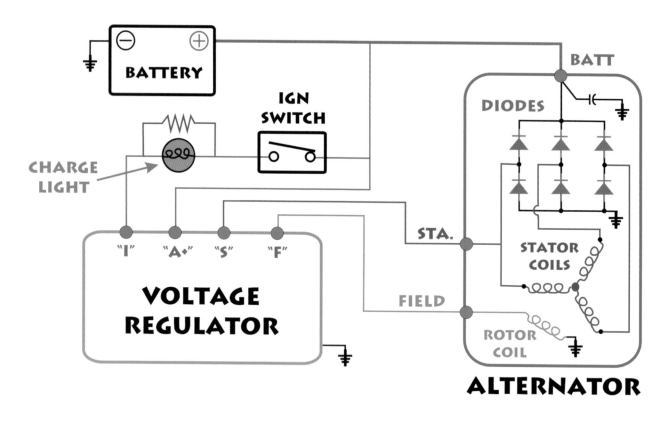

Fig 5-9. *A charge light indicator receives power from the ignition switch and ground from the voltage regulator. When the alternator starts turning, the ground for the voltage regulator's light builds voltage until the indicator light eventually has no ground and goes out.*

V-belts need to be a lot tighter than most people think. If the belt moves easily when pushed by hand, then it's too loose. As a result, the alternator may not charge to full capacity.

Serpentine belts have replaced old V-type belts. Serpentine belts are more reliable and can transfer more torque between an engine's crankshaft and power-driven accessories, like an alternator or air conditioning compressor.

Always read the warning label! This label states: "Battery must be fully charged and in good condition before this unit is installed or serious damage to the alternator could occur." Make sure to charge a dead battery before trying to start a car with a newly installed alternator.

engine speed and alternator speed increase. This charge light indicator flaw plagues older and newer vehicles alike.

ALTERNATOR TESTING

Like most people, you've probably experienced any or all of the following problems with a vehicle. The alternator idiot light is on. The voltmeter on the dash is reading in the no charge zone. The battery has died overnight. All of these symptoms are most likely caused by a charging system that is not doing its job. It happens more frequently than you might think and plagues amateur and professional technicians alike. In fact, charging system problems are, more often than not, caused by something stupid rather than as a result of a component malfunction. As the most likely sources of alternator malfunction, the following common problems need to be eliminated first in order to prevent needless hours of electrical testing. They are listed in order of the most common problems first: (1) belt tension, (2) battery condition, and (3) wires, fuses, connectors, and fusible links.

Belt Tension

By far the most common, no-charge problem is a loose alternator belt. This is less of a problem on newer cars using

This 16-volt battery provides extra voltage to start modified engines. The extra voltage provides more push at the starter dealing with a high-compression engine. This battery also features a manifold venting system designed to prevent acid spillage. Courtesy of New Castle Battery and Summit Racing Equipment

serpentine belts equipped with spring-loaded belt tensioners. Serpentine belts are flat with several V grooves, so they have more surface area than standard V-belts. They can get cracked or glazed over time and should be checked for problems before condemning an entire charging system as the cause of a dead battery. When bad, serpentine belts develop missing string-like sections, a condition that can cause them to slip.

However, an older vehicle with a V-belt is a different story altogether. This type of belt is tensioned during installation, and as a result, needs to be checked periodically for cracks, tears, splits, and glazing—which appears as shiny sections along the length of the belt. Alternator belts should be adjusted much tighter than most people think. If a belt can be easily moved up and down, it's too loose. It often takes a long screwdriver or pry bar to put enough tension on a belt before tightening the alternator hold-down bolt. Also, don't overlook possible loose alternator mounting fasteners as a cause for belt slippage. In addition, the presence of multiple belts can fool you into thinking the alternator belt is tight. For example, a 1988 Cadillac Fleetwood Brougham uses three belts to drive the alternator. The first drives the power steering pump, which in turn uses the second belt to drive the water pump, which then uses yet a third belt to drive the alternator. On these vehicles, even when the alternator belt is tight, if either of the other two is loose, the alternator won't charge the battery during high electrical loads.

Battery Condition

If a vehicle battery keeps going dead overnight—or worse, after only several hours—replacing the alternator won't necessarily fix the problem. This may seem obvious at first glance, but the failure to test a battery first is the cause of far too many

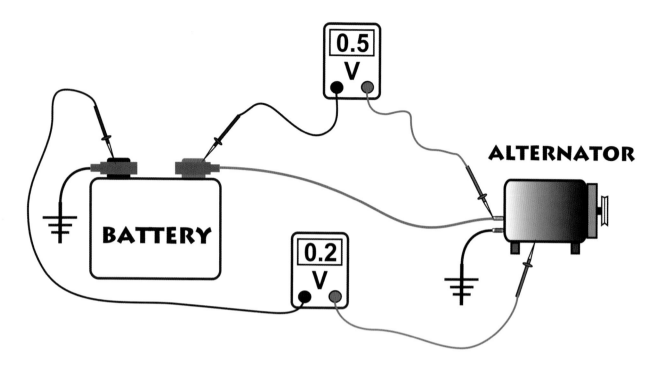

Fig 5-10. *This alternator is putting out 80 amps at 2,000 rpm. The drop of 0.5 volts between the alternator's output terminal and the battery is acceptable for this application. The ground side voltage drop is 0.2 volt, so the ground return is good.*

needless alternator replacements. Even if the battery is fairly new, it doesn't necessarily mean it's good. Test it to make sure, and don't forget to check for loose battery cable connections while you're at it. (See Chapter 4 for battery testing procedures.) Don't forget, the battery should be fully charged before testing and/or installing a new alternator. Jump starting a car with a dead battery and newly installed alternator may damage the alternator as a result of the initial heavy electrical load placed upon it. Furthermore, if the battery is really dead (less than 5 volts of open circuit voltage), the alternator may not recognize it as part of the circuit, since the end-result is the same as if the battery was disconnected from the vehicle. The alternator must balance its voltage output against the battery's internal resistance. As there is no battery seemingly present in the electrical system to push against, alternator voltage increases in excess of possibly 20 volts. With 20 volts coming out of the alternator, other electrical components such as onboard computers and ignition modules may not survive. Do not disconnect the battery on any vehicle equipped with an alternator while it's running; it will put a severely large dent in your wallet! (In fact, disconnecting the battery with the engine running was even a questionable test for DC generators. But since those vehicles didn't have solid state electronics—which are easily damaged by such a procedure—it wasn't quite as risky.)

Wires, Fuses, Connectors, and Fusible Links

Always make sure to check for loose wires or connectors at the alternator, voltage regulator, and battery before assuming an alternator is faulty. Look for connectors not plugged in to the alternator, since you never know who might have forgotten to plug them in during installation. Also, just because a connector may appear tight and not feel loose, this doesn't mean it's actually working electrically. A voltage drop test is the fastest way to determine the strength and solidity of the connection between the big wire at the back of the alternator and the positive battery terminal. It will also help you ascertain if the voltage produced at the alternator is actually getting to the battery. (See Chapter 2 for voltage drop testing.)

To perform a voltage drop test on an alternator, connect the red lead of a voltmeter to the alternator's battery terminal and the black lead to the positive terminal of the battery. Start the engine. Hold engine speed at 2,000 rpm and turn all electrical loads on. The voltage drop should measure from 0.2 volt for alternators with voltage output capacity of around 50 amps, to 0.7 volt for alternators with output capacity closer to 100 amps. (The greater the amperage, the greater the range of acceptable voltage drop.) If it's higher than it should be, then there's a problem with the big alter-

A Fluke i410 AC/DC current clamp is clamped around the alternator's output wire. With the engine running at 2,000 rpm and the amp clamp connected to a multimeter that reads milli-volts (since this current clamp's output is in millivolts), the amperage reading should be within 10 percent of the alternator's rated voltage output. Courtesy of Fluke Corporation

nator wire or its connections. If this is the case, perform the same test on the ground side of the alternator. Connect the red voltmeter lead to the alternator's case and the black lead to the negative battery terminal. Voltage drop should not exceed 0.2 volts. If the voltage reading is higher, there's probably a bad connection at the point where the negative battery cable attaches to the engine block. Also, some alternators use an external ground wire that can cause a poor ground return connection as well.

Fusible links can also cause charging system problems. These circuit protection devices can be hidden in various places under the hood. Once you locate them, if they look burned or the insulation is hard or cracked, they need to be replaced even though they may still work. Don't overlook fuses under the dash or hood when searching for potential causes of charging system problems either. Some vehicles use a charge fuse to power the voltage regulator. If it's burned out, the alternator won't charge. On vehicles designed with engine management systems for controlling the charging system, make sure the CHECK ENGINE light is not on. If it is, check the trouble codes to see if any are charging system related.

GENERIC ALTERNATOR TESTING

The first step before testing an alternator is to ascertain how much electrical load the alternator will be required to keep up with. Connect an inductive ammeter probe to the positive battery cable. Switch the headlights on and check if the meter's reading is a negative or positive number. If it's negative, the ammeter is connected correctly; if it's positive, unclamp the probe, turn it 180 degrees, and reconnect it. With the ignition key turned to ON, but the engine *not* running, turn all electrical loads on, including headlights and high beams, heater blower (on high speed), brake lights, stereo, cigarette lighter, and all other electrical accessories. Record the highest amperage reading displayed on the meter; this is the total amount of amperage from all electrical loads. The meter will display a negative number because it's measuring current from the battery. Once the engine is started, the alternator must produce the same number of (positive) amps in order to break even. However, the alternator must produce an additional 5 amps to keep the battery charged. This number represents the amount of amperage the alternator must be able to produce in order to both power the accessories on the vehicle and keep the battery charged.

Once you've performed the test, shut off all the loads and start up the engine. Hold the engine rpm steady at 2,000 and turn all the electrical loads back on. If the charging system is functioning as it should, the ammeter should display a positive number, confirming if the alternator is providing enough charging current for all accessories (the

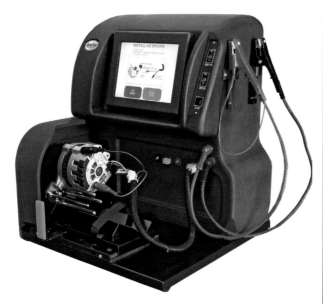

Northwest Regulator manufactures industrial-grade alternator testers used by many auto parts stores. As a last resort, you can always take your questionable alternator in for a test. Courtesy of Northwest Regulator

negative number recorded earlier) as well as the battery. Add whatever positive number is displayed to the negative number recorded earlier. This figure should be within 10 percent of the rated output capacity for the alternator.

By way of example, let's assume that all electrical loads in a vehicle total 52 amps as measured from the battery (meter reads 52 amps). With the engine running and all loads turned on, the ammeter should read positive 5 amps (+5 amps). This number is the aggregate net effect of the alternator's output upon the system (which has counteracted the battery's negative amperage output). The math looks like this: -52 amps (from battery) + 52 amps (from alternator) + 5 (additional) amps (from alternator to power the battery) = +5 amps (displayed on meter). (More simply: -52 + 52 = 0 +5 = 5). It's important to remember the total amperage output coming from the alternator itself is actually 57 amps. The math for the alternator's total amperage output looks like this: 52 amps (to match the total output from the battery for all the loads) + 5 (additional) amps (to power the battery) = 57 total amps. Thus, if a charging system is equipped with an alternator rated at 60 amps, the 57 total amps the alternator produces is within the acceptable 10 percent range for the alternator's rated voltage output capacity.

In addition to checking charging amperage, be sure to check charging voltage as well. Make sure to perform this test with all the loads on and the engine holding steady at 2,000 rpm. Alternator output voltage should be at least 1 volt over the battery's open-circuit voltage—usually somewhere between 13 and 15.5 volts. If the only testing equipment available is a voltmeter, only charging voltage can be read. As long as charging voltage is above open-circuit battery voltage, the voltmeter will confirm the alternator is producing some current, but not how much. Consequently, this test is inconclusive. Unfortunately, without the ability to read amperage, you can only guess (and hope!) that the alternator is okay.

In addition to converting AC voltage into DC, diodes inside the alternator will prevent the battery from discharging back through the alternator when the engine is not running. By indirectly measuring an alternator's maximum output, you can test if any of the six diodes inside the alternator is failing to do its job, since one bad diode can steal about 4 amps from a battery and cause it to go dead in a few hours. This phenomenon is known as parasitic amp draw. (More information on this symptom can be found in Chapter 9 on troubleshooting electrical systems.) To test for this condition, clamp an ammeter probe around the big wire going from the alternator to the positive battery terminal. With the ignition key in the OFF position and all accessories turned off, no amperage should be present in this wire. If the ammeter indicates over 1 amp, then the alternator has one or more bad diodes.

Another diode test uses a voltmeter to measure AC voltage that is leaking past a diode about to go bad. To perform the test, connect the red voltmeter lead directly to the back of the alternator's output wire. Set the scale on the voltmeter to read AC millivolts. AC voltage should not exceed 55 millivolts AC with the engine running and several accessories turned on. If AC voltage is greater than 55 millivolts, then one or more diodes may be close to malfunctioning.

FULL FIELDING AN ALTERNATOR

A full field test will force an alternator to produce maximum current. When full fielding occurs, an alternator's internal or external voltage regulator is bypassed, causing full alternator output. By providing full battery power to an alternator's field coil via a full field test, the alternator will charge at full capacity, proving the alternator is not the cause of any charging-related problems. If the alternator is capable of producing full output during full fielding, the voltage regulator may be to blame for charging-system problems. However, before replacing an external regulator, all the wires connected to it need to be checked with a voltmeter to ensure they have correct electrical values. If an internal voltage regulator is used and the alternator passes a full field test, you may want to consider replacing everything at once, because if one component is worn out, the others can't be far behind.

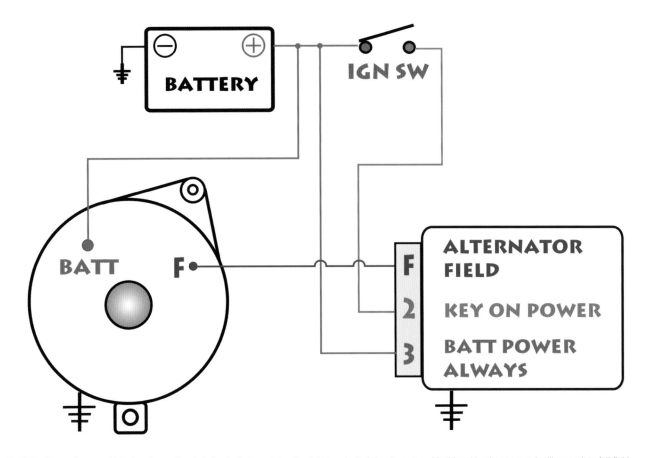

Fig 5-11. *Depending on which charging system is being tested, providing the field terminal of the alternator with either 12 volts or ground will cause it to full field and produce maximum amperage.*

Full field test results are typically similar for most vehicles, but the way the test is performed differs according to the year, make, and model of the vehicle. Some general guidelines for full fielding are included here, but to know for certain which wires to jump, power, or ground, you should consult the service manual for the particular vehicle. In every test, the engine should be held at 2,000 rpm and an inductive ammeter probe should be connected to the big wire(s) at the back of the alternator. Once the alternator has been full fielded, the engine rpm will drop and the alternator will make a whining sound. The ammeter reading should then increase until it reaches within 10 percent of the alternator's rated output. Be sure to full field an alternator for the shortest possible amount of time, usually only about three seconds, to get an ammeter reading. Full fielding longer than this can overheat the alternator, causing damage.

GENERIC FULL FIELD TEST

This test works on most alternators equipped with a separate field wire. Remove the field wire from the alternator and connect a test light to the field terminal. With the key in the ON position, touch the pointed end of the test light to the positive battery terminal. If it lights up, then the alternator's field terminal needs 12 volts to perform a full field test. If the test light doesn't shine while touching the positive battery terminal, but does once the negative battery terminal is touched, then the alternator's field wire needs to be grounded to perform a full field test. If the test light doesn't light up when touching either the negative or positive battery terminal, then the alternator's field circuit is open or a wire in the field circuit is disconnected. Consult a wiring diagram to determine where the field circuit power originates. This problem must be repaired before further testing can be performed.

GM ALTERNATOR WITH EXTERNAL REGULATOR

The charging system for a GM alternator with external voltage regulator looks like this: The alternator has two wires, labeled BATT and F. The BATT wire comes directly from the battery and the F wire is the field wire from the voltage regulator. The regulator has three wires, labeled F, 2, and 3. The F terminal is the field control wire for the alternator; the 2 wire is key's on power; and the 3 wire goes to the battery (see Figure 5-11 on page 85).

If this charging system isn't charging, the first step in testing is to check the wires at the voltage regulator. To do this, remove the regulator connector and turn the ignition key to ON. Terminals 2 and 3 (in Figure 5-11) should light up a grounded test light. Using an ohmmeter, test the F terminal for continuity to ground. If this wire is open, a wire on the field coil inside the alternator is open. If all wires at the voltage regulator connector don't have proper electrical values, the charging system won't work. Consult a wiring diagram to determine the originating power source for each wire. If all the regulator wires are okay, a full field test of the alternator is next.

To full field test this alternator, run the engine at 2,000 rpm and connect one end of a jumper wire to the big wire at the back of the alternator. Touch the other end to the field terminal on the back of the alternator. The alternator

This Delco Remy alternator features a test hole at the back of the alternator. Just stick a pocket screwdriver or Allen wrench into this hole and the alternator will full field. Courtesy of CARQUEST Auto Parts

Fig 5-12. The BATT and F2 wires depicted here should have battery voltage. F1 comes from the charge indicator light. All of these wires must have correct electrical values before an alternator will work. The back of the alternator's case has a small hole labeled "test hole." A small screwdriver inserted into this hole will full field the alternator.

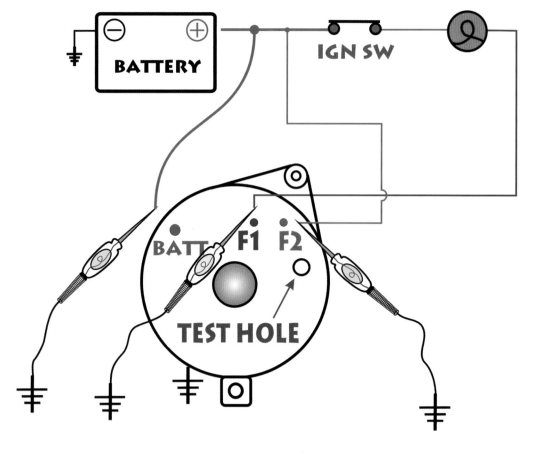

GM produces two CS alternators that can be identified by the letters on the plastic plugs at the backs of the cases. Each plug will read either PLIS or PLFS. These alternators are not interchangeable, so make sure the correct one for the vehicle is installed. Courtesy CARQUEST Auto Parts

should full field and produce maximum current. If alternator output is less than expected, get a new alternator.

GM DELCOTRON ALTERNATOR WITH INTERNAL REGULATOR

GM vehicles are also commonly equipped with a Delcotron alternator (also known as a Delco-Remy 10-S1). This charging system uses an internal voltage regulator and is easily identified by the presence of three wires connected to the back of its case. The two wires, BATT and F2, go directly to the battery. The F1 wire goes to the charge indicator light on the dash (See Figure 5-12).

The first step in testing this charging system is to verify that the charge indicator dash light works. If the indicator light doesn't come on when the ignition key is in the ON position, the alternator won't charge. If this is the case, the charging light circuit needs to be repaired before further testing.

To test terminals F1 and F2, unplug the connector from the back of the alternator. Using a grounded test light, touch each of the three wires, one at a time. The big wire should light up the test light brightly. One of the smaller wires (labeled F2 on the diagram depicted in Figure 5-12) should also brightly light the test light. F1 should light the test light, too, but only dimly. If any of the wires don't light the test light as described, the alternator won't charge and the circuit must be repaired before proceeding further.

If all the wires of the alternator have correct electrical values, a full field test is the next step. Plug the connector back into the alternator and start up the engine. Hold the engine at 2,000 rpm and insert a small screwdriver or Allen wrench into the regulator grounding hole (test hole), located at the back of the alternator. Doing this will full field the alternator, causing it (hopefully!) to produce full amperage (within 10 amps of its rated output capacity); if alternator output turns out less than expected, then it's time to shop for a new alternator.

DELCO-REMY CS ALTERNATOR

Delco-Remy CS alternators come in two flavors—PLIS and PLFS. They can be identified by reading the letters on the plastic plug on the back of the alternator. PLIS alternators are usually (but not always) found on vehicles without a body control module (BCM), while vehicles with PLFS alternators usually have a BCM. These two types of alternators are *not* interchangeable, which can be a problem when purchasing a rebuilt alternator. In addition to checking the letters on the plug, make sure you have the correct part number to match your car or truck. Delco-Remy CS alternators use a charge indicator lamp that must be operational in order for the charging system to function. If the lamp does not light up when the ignition key is turned to ON, disconnect the plug from the back of the alternator and ground the L terminal; the lamp should now shine. If it doesn't, check for an opening in the lamp circuit or a burned out indicator bulb.

The normal system voltage produced by CS alternators is somewhat lower than on other charging systems. A typical range of readings, taken at the battery's positive terminal with the key turned to ON and engine off is 12.3 to 12.5 volts. With the engine running at idle, the range should be between 12.8 and 13.1 volts. With all accessories turned on, the range is between 12.8 and 13.3 volts. The internal voltage regulator is designed to change the charging voltage slowly, so as to help the ECM keep engine-idling speed stable.

A Delco-Remy CS alternator cannot be full fielded; instead, use the generic alternator test previously mentioned to check its current and voltage outputs. If the alternator cannot produce close to rated amperage, the wires at the alternator plug need to be checked for correct electrical values. The following are the electrical values for each of the wires for each type of CS alternator plug:

• The **P** wire is an AC voltage signal used for tachometer input (it is not needed for charging).

• The **L** wire sends a signal to the BCM if the regulator is working. Obviously this only occurs on a BCM-equipped vehicle; on non-BCM-equipped vehicles, the L wire lights the charge indicator lamp and powers the regulator.

• The **F** wire (again, only on a non-BCM-equipped vehicle) carries the power (with the key turned to ON) used to turn on the voltage regulator. If the vehicle has a BCM, the F wire informs the BCM how hard the alternator is working. The internal regulator switches the alternator's field coil on and off at about 400 cycles per second—the amount of on time determines charging output. At high engine speeds and with low current requirements, the on time may only be 10 percent. However, at idle or at low engine speeds with high loads, the on time can be as high as 90 percent. The signal can be read using a DVOM capable of measuring frequency.

• The **I** wire is used only on non-BCM-equipped vehicles and receives 12 volts from the ignition switch (or sometimes through a resistor).

• Last but not least, the **S** wire senses battery voltage for both types of alternators.

Shown is the plastic plug on a Delco-Remy alternator. This plug will be marked PLFS or PLIS. These alternators are not interchangeable, so make sure you have the right one for your vehicle before installing.

DASH AMMETER

TO ALT.

F
S
A+
I

FIELD COIL = CONTINUITY TO GROUND (CHECK USING OHMMETER)

BATTERY VOLTAGE W/IGNITION KEY ON

BATTERY VOLTAGE ALL THE TIME

NOT USED WITH DASH AMMETER

IDIOT LIGHT

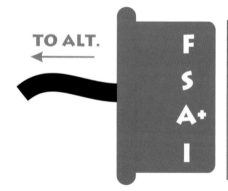

TO ALT.

F
S
A+
I

FIELD COIL = CONTINUITY TO GROUND (CHECK USING OHMMETER)

STATOR = 6/7 VOLTS - "F" AND "A+" JUMPED, ENGINE RUNNING

BATTERY VOLTAGE ALL THE TIME

CHARGE INDICATOR LAMP = BATTERY VOLTAGE W/IGNITION KEY ON

Fig 5-13. *Full fielding is accomplished in the same manner for both types of Ford's charging systems. To full field, jump terminals F through A+. The alternator should produce amperage within 10 percent of its rated output capacity.*

This Ford IAR alternator has two plugs. The plug on the back of the alternator (black colored, upper left) is the voltage regulator. The brown plug on the side of the alternator is for high-amperage output wires going directly to the battery. Courtesy of CARQUEST Auto Parts

If all the wires at the plastic plug have correct voltage readings and the CS alternator fails the generic alternator test, the answer to your charging problem is a new alternator.

FORD ALTERNATOR WITH EXTERNAL REGULATOR

Ford vehicles equipped with alternators controlled by external voltage regulators use different pairs of alternators and regulators that are not interchangeable. One regulator/alternator pair is in vehicles with an ammeter on the dash; the other is in vehicles with a charger indicator light. Both regulators have four terminals, labeled F, S, A+, and I, and both alternators have four wire connectors that plug into the regulators (see Figure 5-13).

To full field either alternator, unplug the four-wire connector from the regulator, start the engine, and maintain rpm at 2,000. Using a jumper wire, connect the regulator harness of terminals F and A+ together. The alternator should full field and produce maximum output. If it doesn't, there may be a problem with the wires between the alternator and regulator connector, so the same test should be performed again, but this time try connecting the battery and F terminal together at the back of the alternator. If the alternator full fields now, the wires going to

the regulator connector must be repaired. (See Figure 5-13 to identify terminals on either regulator.) If the wires to the regulator have correct electrical values, but the alternator does not produce amperage within 10 percent of its rated output during a full field test, it's time for a new alternator.

FORD-MOTORCRAFT IAR ALTERNATORS

Ford-Motorcraft IAR alternators use internal voltage regulators. These alternators can be full fielded by grounding the F terminal located on the regulator at the back of the alternator. If the alternator doesn't produce full current output, check the connector(s) plugged into the alternator. Ford alternators with outputs of 60 amps usually have one three-wire connector; higher output units may have two three-wire connectors.

To test an alternator equipped with only one three-wire connecter, unplug the connector and turn the ignition key to ON. Using a voltmeter, check the terminals at the connector as follows: terminal I (connected to the charge light indicator), terminal S (connected directly to the battery), and terminal A (connected to the ignition switch—it receives key on power, which it sends to the regulator). All of these should read 12 volts.

This Ford IAR alternator uses an internal voltage regulator. "Ground here to test" is marked on the regulator. To full field the alternator the F terminal on the regulator is grounded.

To test an alternator with two three-wire connectors, use the following procedure. Test the first connector (plugged into the voltage regulator) in the same manner as above. The second three-wire connector (plugged into the side of the alternator) has two big wires leading directly to the battery, and a smaller one that loops back into the first connector's S terminal (also connected to the battery). All three wires at this connector should have battery voltage all the time. If the connector(s) have correct electrical values and the alternator doesn't produce maximum amperage when the full field test is performed, either the alternator or regulator is bad. In such a case, both the alternator and regulator should be replaced together, because if one of these components has gone bad due to old age, vibration, or heat, the other won't be far behind.

CHRYSLER ALTERNATOR WITH EXTERNAL VOLTAGE REGULATOR

Chrysler also manufactured vehicles equipped with alternators fitted with external voltage regulators. This alternator is readily identified by its three-wire connector.

The biggest wire connects to the positive battery terminal, while the two smaller wires are for the alternator's field coil. (Some of these alternators may have a battery ground wire.) To full field this type of alternator, you must first identify the field terminal at the regulator. Unplug the voltage regulator and turn the key to ON. Using a grounded test light, probe both wires going to the alternator. One wire should light the test light up brightly. The

other should either make the test light dimly light up, or not at all; this wire (usually green) is ordinarily the ground used to full field the alternator. Start the engine and hold rpm at 2,000. Using a jumper wire, momentarily connect the field wire to ground to full field the alternator. If the alternator output has correct amperage, it's time for a new voltage regulator. If the output isn't at maximum, try a full field test again, but this time perform the test at the back of the alternator. If it full fields now, simply repair the broken field wire. If it still won't full field, there's no other option—it's definitely time for a new alternator.

CHRYSLER'S COMPUTER-CONTROLLED ALTERNATOR

Chrysler's computer-controlled alternators are similar to those alternators using external voltage regulators. Look for two small wires (usually green) at the back of the alternator. One of these wires comes from the ignition switch and should have 12 volts when the key is turned to ON. The other comes from the power module and controls the ground of the alternator's field coil and, thus, alternator output. To test this type of alternator, first turn the ignition key to ON. Using a grounded test light, touch each field terminal at the back of the alternator. One terminal will make the test light shine brightly, while the other will make it light dimly or not at all. The latter field wire is the one that will need to be grounded to full field the alternator. Start the engine and hold rpm at 2,000. Using a grounded jumper wire, momentarily touch the field wire to ground. The alternator should full field and produce maximum current. If it does, but the charging system still isn't charging the battery, the problem is most likely at the power module. Because Chrysler power modules are expensive, always consult a wiring diagram *before* replacing this part in order to ensure all power and ground wires into the power module are good. In addition, refer to the service manual, as it tells you how to read charging system fault codes. These should also be checked before replacing the power module. Chrysler's late model vehicles now use an integrated single computer (instead of dual computers) for controlling charging system functions. Fortunately, this alternator operates in a manner similar to earlier dual-computer charging systems and, thus, can be tested in the same manner with the same expected results.

DC GENERATOR AND REGULATOR TESTING

Even though DC generators are old school, there are numerous collectors and vintage automobile enthusiasts who will, sooner or later, have to test these systems. Just like modern charging systems, old DC generators with voltage regulators can be full field tested by following the same general procedures. Voltage regulators used in conjunction with DC generators have only three wires. The terminals are typically labeled F (which goes to generator's field wire), BATT (which goes to the battery's positive terminal), and A (which goes to the armature in the generator). Just like an alternator, a DC generator can be full fielded to test its maximum output. To full field a generator, start the engine and hold it at 1,500 rpm. Disconnect the F terminal at the voltage regulator and touch the wire to the positive battery terminal. The generator should full field. (On some Chrysler models, touching the F wire to ground will full field the generator instead.) If it still won't full field, try doing the same test directly at the generator. If it full fields now, repair the field wire going to the regulator; if it doesn't, try polarizing the generator (discussed in next section) before getting a new one. Be sure to check the three wires at the regulator and generator to be certain they're connected properly. Consult a wiring diagram to identify all wires.

If the voltage regulator's wires are okay and the generator full fields, then the voltage regulator is probably bad. Before purchasing a new regulator, make sure it's the right one for the vehicle. Many older vehicles have positive ground systems. Also, pay attention to the regulator's working voltage—6 or 12 volts, depending upon application. In some instances, you may be able to use an electronic regulator as a better replacement for a mechanical one. And don't forget, some vehicles use an in-dash series ammeter, and if that is not working, the battery won't charge, even if the generator and regulator are good.

GENERATOR POLARIZING

If a new or rebuilt generator is installed into a vehicle, it must be polarized before it can produce current. Polarizing establishes correct polarity for the magnets inside the generator.

Here is a general procedure for DC generator polarizing. Disconnect the field wire at the voltage regulator. Connect a test light to the field wire and touch the end to battery positive. If the test light shines, remove it and momentarily touch the field wire directly to the battery terminal on the regulator to polarize the generator. When the field wire is touched to the BATT terminal, a good-sized spark should be produced; this is normal. However, if the test light does not light up when touching battery positive, touch it to the negative battery terminal instead. If it now lights up, reconnect the field wire to the regulator and remove the armature (A) wire from the regulator. Momentarily touch the armature wire to the battery terminal on the regulator to polarize the generator. When you touch the armature wire to the BATT terminal, a good-sized spark should be produced. Again, this is normal. However, if the test light fails to light after touching either the positive or negative terminal of the

This Toyota starter motor uses a gear-reduction set to increase starter torque. This design allows for a smaller overall starter size to be used to produce the same or greater torque than that found in larger units. Courtesy of Younger Toyota

A starter solenoid has two jobs: connect the battery directly to the starter and engage the starter pinion gear with the engine flywheel. Courtesy of Younger Toyota

This aftermarket starter provides 25 percent more torque than the original starter it replaced. The starter shaft is heat-treated and rotated in Bendix bearings that provide reliability and long life. Courtesy of Summit Racing Equipment

battery, the field circuit inside the generator is open and needs to be repaired. This may involve finding a break in the wire and replacing it with a new section of wire.

STARTER MOTORS

A DC starter motor operates much like a DC generator—only in reverse. It uses current to operate, instead of producing current when it spins. However, just like a generator, a starter motor uses field coils to create a magnetic field around an armature, which in this case is a series of wire loops connected at a commutator. When the starter motor is operating, current from the battery energizes the field coils, causing them to have a strong magnetic field. At the same time, battery current is applied to the commutator brushes (that carry current from the battery to the armature). The armature rotates because the opposing magnetic lines of force between the field coils and armature repel each other, and since the armature has multiple wire loops, it rotates continuously. Starter motors may use as many as four commutator brushes and from two to four field coils.

The armature and field coils are in series, a design which causes a very high torque output from the starter. With the field windings and armature in series, any increase in current produces an increase in field strength. As the load on the starter increases, so does the starter's torque. Under a no load condition, such as during starter bench testing or when a starter's pinion gear doesn't engage the flywheel, the starter speed will continue to increase until centrifugal force destroys the starter. It then makes a loud "bang," followed by a shower of sparks.

Certain starter motors use gear sets to increase cranking speed and torque. A planetary gear set (similar to those used in automatic transmissions) sends power between the starter motor and output shaft connected to the engine flywheel. Newer starter motors use permanent magnets instead of field coils which increases current going to the armature.

Using an inductive amp probe, like this Fluke i410, will confirm whether a starter's current draw is too high or too low. This test is faster and easier than replacing a starter—only to find out later that some other component was the real culprit. Courtesy of Fluke Corporation

STARTER SOLENOIDS

Starter solenoids perform one, and sometimes two, functions; they connect the battery directly to the starter, and on some starters, also engage the pinion gear with the engine flywheel. Solenoids can be mounted directly onto the starter case or used remotely. They are really just overgrown relays that use a small amount of current to energize a coil of wire to produce a magnetic field. The strength of the magnetic field pulls the solenoid's plunger into contact with two terminals—one from battery positive and the other to the starter. Positive engagement starters use a shift fork, or a lever, that is connected to the solenoid mounted directly onto the starter. When the starter receives battery voltage, the solenoid moves the lever, causing the starter to engage the engine's flywheel. Starters using remote solenoids have moveable pole shoes that move a yoke when the starter is energized. The yoke pushes the drive gear into the flywheel in order to crank the engine.

STARTER TESTING

Before testing a starter motor, the vehicle's battery must be fully charged and load tested. There's simply no point in trying to figure out which part of a starter circuit is bad if the battery doesn't have enough energy to crank it, even when nothing's wrong. (See Chapter 4, Storage Battery, for testing and/or charging procedures.) The only real reason to test a starter is when it's cranking an engine too slowly to start it or not cranking at all. In certain front-wheel-drive cars, the starter motor is difficult to remove (to say the least!). By contrast, performing basic starter testing takes only a few minutes and is a much more productive use of time than spending six hours changing out a starter, only to find the problem is really a bad battery cable. In fact, rather than testing a starter, it's easier to test everything else first. If the wires, cables, connections, and solenoid are good, the starter is the only thing left to replace.

The test that provides the most information with the least amount of work is a starter current-draw test. To perform this test, connect an inductive ammeter to the positive battery cable, turn the headlights on, and read the ammeter. If it reads negative, the probe is connected correctly; if it reads positive, turn the amp probe 180 degrees and reconnect it. Don't forget to turn off the headlights. Then get ready for the next step—disabling the ignition system.

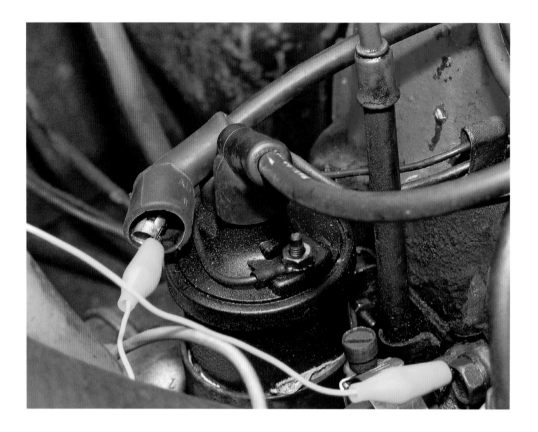

This high voltage coil wire has been grounded using a jumper wire. This prevents the engine from starting up when testing a starter motor.

Disabling the Ignition System

The next step is to prevent the ignition or fuel injection system from starting the engine when the starter motor is cranked. On older cars with only one ignition coil, find the wire going from the distributor cap to the coil, remove it from the distributor cap, and connect a jumper wire. Connect the other end of the jumper to ground; this will prevent the engine from starting because the spark from the coil bypasses the spark plugs and goes directly to ground. By simply unplugging the ignition coil (without grounding it first) and cranking the engine, the ignition module or other parts can be wiped out. On later model vehicles without distributors, to disable the ignition system, simply unplug the connector going to the ignition module or locate and remove any fuse(s) that power(s) the computer or ignition system.

Normal Starter Amperage Draw

Before abnormal starter amp draw can be recognized, you must first know what normal looks like. All of the following specifications for starter draw assume an ambient temperature above 60 degrees Fahrenheit and a conventional starter motor cranking an engine. A four-cylinder engine should draw between 50 and 125 amps; a six-cylinder and a small V-8 require between 75 and 175 amps; and a large V-8 engine draws in the range of 100 to 275 amps. Some

service manuals provide more specific numbers, but these work most of the time for testing purposes. Gear-reduction and permanent magnet type starters yield slightly different results; be sure to check their service manuals for starter draw specifications.

Slow Turning Starter—High Amperage

Although a slow-turning starter with high amperage is not commonly encountered, it does occur occasionally, so follow the next procedure to diagnose this problem. Turn the ignition key to START position, then crank the engine over (do this even if the starter won't crank at all) and check the ammeter reading. If the amp reading is excessively high and the engine is turning too slowly or not at all, there may be a problem with the starter motor or engine. To eliminate the engine as the cause of a slow-turning starter, turn the crankshaft by hand using a breaker bar and socket. On a four-cylinder engine, try grabbing the alternator/water pump drive-belt as a means of turning it. Even with spark plugs installed, the engine should rotate with relatively little force. If it takes both feet and arms to move the engine minimally, a bad starter is the least of your problems. The engine could have a mechanical problem like high-viscosity oil in weather colder than 20 degrees Fahrenheit or carbon build-up in the cylinders. Another possibility is that there is no oil and the

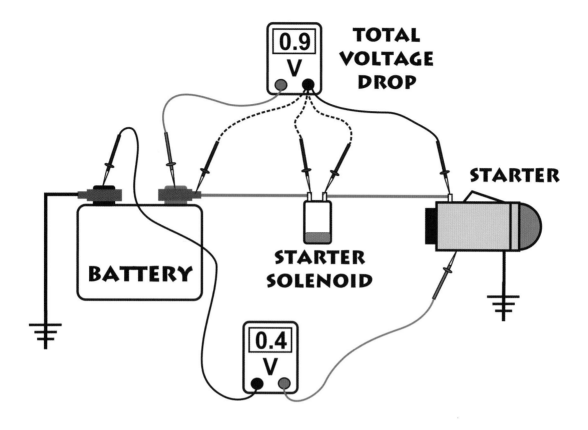

Fig 5-14. *Ground side resistance in this starter circuit is okay because the voltage drop is only 0.4 volt. The positive side also has a voltage drop of 0.9 volt caused by high resistance as well. By moving the black lead along the positive side of the circuit, the point of high resistance can be located.*

crankshaft has welded itself onto the connecting rods or overheated, causing other mechanical problems. Conversely, if the engine turns okay, get a new starter because the flow of high amperage into the starter circuit indicates a shorted starter armature or field coil—not an engine problem.

Slow Turning Starter—Low Amperage

If, on the other hand, the ammeter reading is low and the starter turns slowly or not at all, the starter circuit has high resistance that is possibly caused by poor battery cable connections or bad solenoid. High resistance is, by far, the most common reason for low starter circuit amperage. It can't possibly be a battery problem since the battery's already been tested and charged beforehand (right?). Therefore, a voltage drop test is now an ideal test to use to locate the source of unwanted, high resistance in the starter circuit.

The positive side of the starter circuit should be tested first, as it is the most likely place for problems to occur. Connect the red lead of a voltmeter to the positive battery terminal (not the battery cable). Connect the black lead to the starter terminal where the battery cable attaches. Crank the engine over while watching the voltmeter. If the voltage drop is less than 0.5 volt, the positive side of the starter circuit doesn't have high resistance. Do the same test on the ground side of the starter circuit. Connect the red lead to the starter case and the black lead to the negative terminal on the battery, and then crank the engine. The voltage drop should not exceed 0.4 volt. If either the positive or negative side of the starter circuit exceeds maximum voltage drop (see chart under section How Much Is Too Much? in Chapter 2), the point of high resistance can be located by performing a voltage drop test across individual connections. This can be accomplished by disconnecting the black lead and moving it back along the circuit toward the red lead. There should be no more than 0.2 volt lost across any connection. Cleaning a bad connection is fortunately all that's required to get the starter to spin faster in most cases.

CHAPTER 6
IGNITION SYSTEMS

Ignition-related problems, whether real or perceived, are the cause of more needless replacement of parts than any other automotive electrical system. Before the advent of fuel injection, many people often thought that most carburetor problems could be fixed by looking inside the distributor. Mechanics and owners alike most often blamed the carburetor for engine stalling, missing, and no starts, and since it could be taken apart and visually inspected, they felt comfortable doing this. They often fixed ignition malfunctions with simple replacement of points, condenser, ignition coils, wires, and spark plugs. In early 1970, this method of repair changed with the introduction of electronic ignition systems (EIS), since they were far more costly to replace.

Today, electronic engine management systems control ignition systems on all vehicles. Consequently, it's much more cost-effective to test components before simply throwing them away and replacing them. Modern ignition systems are not as difficult to diagnose as you may think, as there are a great number of similarities between conventional points-type ignition systems and electronic ignition systems. A basic understanding of how primary and

Both of these sets of components produce an arc across a spark plug: points, condenser, and standard coil (left); or an electronic ignition module and high-energy coil (right). Figuring out what's good or bad with each of them is not as hard as you may think.

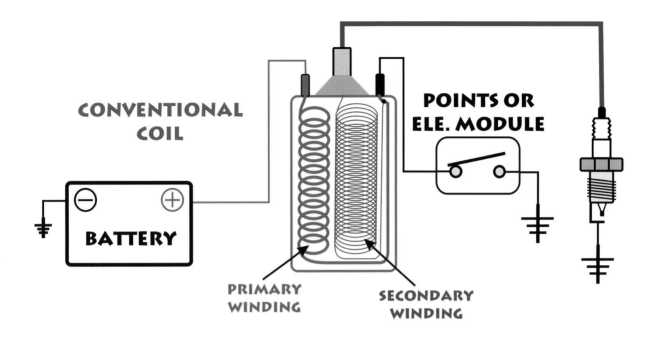

Fig 6-1. *The primary windings actually surround the secondary windings (the drawing separates the two in order to illustrate how they're internally connected). An ignition module or points switch the ground circuit on and off.*

secondary ignition circuits operate goes a long way toward figuring out if you are faced with a bad pick-up coil, hall-effect switch, or electronic ignition module. We'll start by taking a look at conventional ignition systems, which haven't changed much in the past 100 years.

HOW TO GET FROM 12 VOLTS TO 100,000 VOLTS

An ignition system has two jobs—spark production and spark distribution. The latter is accomplished by mechanical means on vehicles with distributors and by electronic means on later model vehicles with distributor-less ignition systems (DIS) or coil-over-plug systems. Spark production on both points-type and electronic systems is basically the same, with the only differences being the internal design of the coils and how they are controlled. Ignition coils operate in an identical manner in both systems and simply increase the electrical voltage from 14.5 volts (charging voltage) to 20,000 volts in points-type ignition systems, and to as much as 100,000 volts in electronic systems. High voltages are required because the high resistance of the air gap on the spark plug serves as the load device in the secondary ignition circuit. Without high voltage, the air gap's resistance cannot be overcome and no spark would occur.

Both conventional and electronic ignition systems have two circuits—primary and secondary. The primary circuit is

the first phase of the process, causing charging voltage to increase from 14 volts to about 400 volts. The circuit is made up of one-half of the ignition coil, points, and a condenser (in older vehicles) or an electronic ignition module or computer (in newer vehicles). The secondary circuit raises voltage a step further, increasing it from 400 volts to over 20,000. The secondary circuit consists of the other half of the ignition coil, the distributor cap and rotor, plug wires, and spark plugs. Thus, every ignition coil is part of both primary and secondary circuits and operates in a similar manner on both points-type and electronic ignition systems.

IGNITION COILS

Ignition coils also have two windings—primary and secondary. The primary winding is made up of about 200 turns of heavy gauge wire. The secondary winding is constructed of lighter wire and more than 20,000 turns of wire wound around an iron core that concentrates the coil's magnetic field. The primary winding surrounds the secondary winding and iron core. When current from the battery flows through the primary winding, a magnetic field is produced inside the coil. The primary circuit is controlled either by a mechanical switch (points) or a transistor in electronic systems. When the primary winding's power is switched off, the magnetic field inside the coil collapses. This collapsing field produces as much as

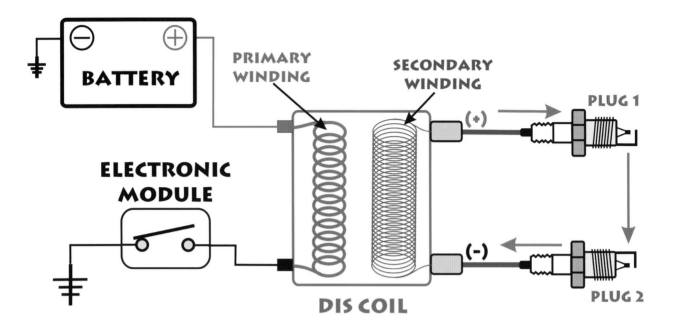

Fig 6-2. *There is no internal connection between the primary and secondary windings on a DIS coil. The high voltage path starts at the positive side of the secondary windings, onto plug 1, through the cylinder head and plug 2, and back to ground (of the secondary windings). Remember: battery ground and the ground for the secondary windings are not the same thing.*

400 volts and induces current into the secondary winding. Because the secondary winding has more turns of wire than the primary winding, voltage is stepped up to as much as 100,000 volts on a late model system.

There are two types of ignition coils: conventional and DIS. Both coil types produce a spark but have different internal circuit designs. Conventional coils are found on vehicles with distributors or coil-over-plug ignition systems. An ignition system with a distributor has a single ignition coil and delivers voltage to spark plugs via a distributor cap, rotor, and high-voltage spark-plug wires. Coil-over-plug systems have individual coils for each cylinder and generally don't have spark plug wires. Conventional coils have three terminals: a power source for the primary circuit, a ground for the primary circuit, and a high-voltage terminal connected to the distributor cap. Power for an ignition coil's primary circuit is supplied from the ignition switch.

A set of mechanical points or a transistor controls (switches on or off) the coil's primary circuit by switching the ignition coil's ground open (not connected to ground) or closed (connected to ground). Switching the ground side of the primary circuit also determines when a spark will be produced.

The coil's primary circuit is connected internally to the secondary circuit at the negative coil terminal. Power for the secondary circuit comes from the primary circuit's collapsing magnetic field, which produces about 300–500 volts. Voltage is stepped up in the secondary circuit, causing current to pass through the high voltage tower in the center of the coil, then onto the spark plug, and finally returning to battery ground (See figure 6-1).

A DIS (distributor-less ignition system) coil is not wired internally in the same manner as a conventional coil. DIS coils have been in use on motorcycles for many years, but when these systems were first introduced for use in cars and trucks in the mid-1980s, they were confusing to technicians who were used to working with a single coil per vehicle.

Two high-voltage terminals are used on a DIS coil (instead of one for a conventional coil), and each coil is connected to a pair of spark plugs. A four-cylinder engine has two DIS coils, a six-cylinder uses three, and an eight-cylinder has four. The primary circuit is powered and controlled in the same manner as in a conventional coil, but the major difference is that the secondary windings are not connected to the primary circuit in a DIS system. The DIS coil's secondary circuit has the same elements as any DC circuit—power, ground, and a load device. One end of the secondary windings is power, the other end is ground, and between them are the load devices (the air gaps of the two spark plugs). When the primary circuit's magnetic field

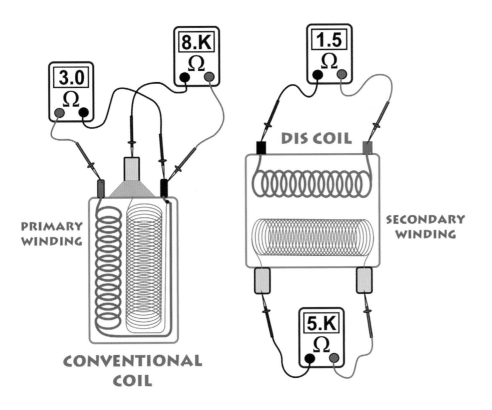

Fig 6-3. An ohmmeter will only tell you if the coil windings are open or shorted. The coil may check out okay, but still not produce a spark. With an ohmmeter connected to the coil try hitting the coil with a screwdriver or heating it with a hair dryer. If any of the resistance readings change, the coil is bad.

8.K Ω

3.0 Ω

1.5 Ω

DIS COIL

PRIMARY WINDING

SECONDARY WINDING

5.K Ω

CONVENTIONAL COIL

collapses, voltage is induced into the secondary windings. Electrons flow from the positive end of the secondary windings to the first spark plug and across the air gap producing a spark. After reaching the first spark plug, current passes through the cylinder head—the equivalent of a really big wire—to the second spark plug's ground electrode, where it jumps the air gap and produces a spark. Each coil pair fires one spark plug in the normal direction and the other plug in reverse. After jumping the air gap of the second plug, the current returns to the other side of the secondary windings—the ground side of that coil (see Figure 6-2).

To make sense of this type of circuit, think of the secondary windings on a DIS coil as having nothing to do with the vehicle's battery ground. Each coil pair fires the spark plugs in two cylinders simultaneously. Both cylinders are at "top dead center" in the crankshaft's rotation. One cylinder is on its compression stroke and the spark ignites the air/fuel mixture. The other companion cylinder is at the end of its exhaust stroke, so the second spark has no effect because there is nothing to burn inside the combustion chamber.

GENERIC COIL TESTING

Both types of coils can be checked for opens or shorts using an ohmmeter. To check resistance on a conventional coil,

The difference between the DIS coil (left) and a conventional coil is that the DIS coil's secondary windings are not connected to the primary circuit. This design causes the spark produced by a DIS coil to go to the other side of the secondary windings and not the battery ground.

This ignition coil is being bench-tested. The yellow wire is 12 volts and the wire coming out of the coil tester goes to ground. When the pointy end of the coil tester is tapped on the negative coil terminal, a spark is produced.

connect one ohmmeter lead to the negative side of the coil and the other to positive. Primary circuit resistance should read between 1.5 and 3.5 ohms. Resistance in the secondary circuit can be measured by connecting one ohmmeter lead to the negative coil terminal and the other to the high-voltage tower; it should be between 7,000 and 15,000 ohms. Remember, the primary and secondary windings are not connected in a DIS coil. Consequently, DIS primary circuit resistance should measure between 0.5 and 2.0 ohms, while secondary circuit resistance should be from 5,000 to 7,000 ohms. A service manual will provide specific resistance values, but the numbers given here are close enough to determine if there is a problem with the coil's windings. However, measuring ignition coil resistance cannot conclusively determine whether a coil is bad, since a coil can check out okay with an ohmmeter but still not produce a spark when electrically loaded (see Figure 6-3).

There are a couple of additional things you can do when attempting to discover if you're dealing with a bad coil. With the ohmmeter connected to the coil's primary or secondary windings, try tapping on the coil with a screwdriver handle or heating it with a hair dryer. If the ohmmeter's readings change, the coil windings are broken.

You can also make an inexpensive universal ignition coil tester out of an old condenser and some jumper wires. Make sure the condenser is good by testing it on a known good coil. To use the tester, disconnect the negative side of the coil and connect the condenser wire to the negative coil terminal. Then ground the condenser's mount tab using a jumper wire. Connect a second jumper wire to ground and turn the ignition key to ON. Now tap the grounded jumper wire to the negative side of the coil; it should produce a spark between the high-voltage terminal and ground on a conventional coil, or between high-voltage terminals on a DIS coil (see Figure 6-4). If you don't get a spark, check for battery voltage at the coil and good connections on all jumper wires. Be particularly careful when doing this on a DIS coil; the spark produced can really zap you if you get in the way. More information regarding coil testers can be found in Chapter 3 on electronic testing tools.

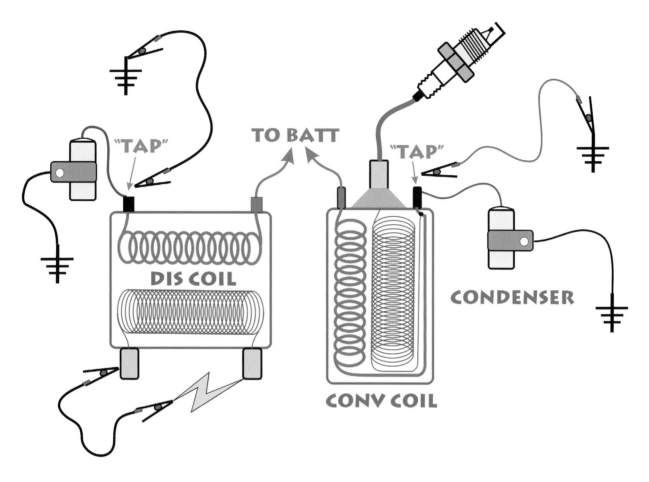

Fig 6-4. *A homemade coil tester will work on both conventional and DIS types of coils. Be careful using it on a DIS coil as the spark can really zap you.*

Conventional ignition coils with threaded or spade terminals can be mistakenly connected backwards, causing reverse polarity and a weak spark. A coil with reverse polarity will start and keep an engine running, but may misfire under load because the coil is producing about 15 percent less voltage than it should. Most coil terminals are marked, but if you're working on one that isn't, you can check coil polarity using a lead pencil. Remove one spark-plug wire and hold it close to the spark plug. Place a lead pencil point in the path of the spark and crank the engine. The spark will flare as it contacts the pencil point. A flare in the direction of the spark plug indicates correct coil polarity; if the spark flares toward the spark plug wire, the coil has reverse polarity and the coil's primary circuit is connected backwards.

POINTS AND CONDENSER

The contact points in a conventional ignition system cause the coil to switch on and off. When the points are closed, ground is supplied to the coil's primary circuit. When the points are open, the primary circuit looses its ground and the magnetic field inside the coil collapses, causing the secondary windings to produce a spark. The points open and close according to engine rpm and are timed via a cam inside the distributor. The closed, or on time, is called ignition dwell. The longer the dwell time, the more the coil gets charged or saturated, creating a larger magnetic field; thus, longer dwell time equals higher resulting secondary voltage. Dwell time is adjustable and can be measured with a dwell meter if you're picky, or by simply measuring the resulting air gap when the points are open. The wider the gap, the smaller the dwell time, and conversely, a smaller gap equals an increased dwell time (see Figure 6-5).

All points-type ignition systems use a condenser as part of the primary circuit. The condenser (or capacitor) is connected across the points and reduces arcing between the contact points as they open and close. Without the condenser, an electric arc would occur at the points instead of at the spark plugs, causing the points to quickly burn out. In addition to the condenser protecting the points, the use of

101

Contact points have been used for many years in automobiles. These simple mechanical switches were subject to wear and had to be adjusted around every 10,000 miles. Consequently, they couldn't provide the reliability needed for the requirements of modern ignition systems.

IGNITION SYSTEMS

an ignition resistor that is in series with the battery and ignition coil limits the current in the primary ignition circuit. By limiting the amount of current capable of reaching the coil, the resistor keeps it from overheating. When the starter motor is operating, battery voltage drops. Oftentimes, a bypass circuit is used in place of the resistor since it can provide more voltage at the coil when the engine is cranked. With the starter drawing down battery voltage, the ignition coil needs a little help in the form of direct battery voltage. Some vehicles use a block-type resistor and others use a resistance wire—both should have resistance of about 1.5 ohms on a 12-volt system.

If you think you may be facing a no-spark problem, pull the high voltage wire out of the distributor and hold it near ground. Crank the engine and watch for a spark to jump between the coil wire and ground. Be aware that if you use only a spark plug wire for this test, it's possible a bad cap, rotor, or plug wire can fool you into thinking you have no spark. A quick way to check the operation of the points is to use a test light connected across the primary circuit of the ignition coil. Connect the alligator clip of the test light to the negative side of the coil and touch the test light's probe to the positive coil terminal. With the test light looped across the coil, crank the engine over. The test light should flash if primary switching is taking place. If the test light

doesn't flash, check the following: (1) power to the coil; (2) the wire between coil negative and the distributor; (3) the points—to see if they are opening/closing; and (4) the distributor—to make sure it's turning. This test will work on both electronic and points-type ignition systems.

ELECTRONIC IGNITION AND COMPUTER CONTROLS

The primary reason for manufacturers switching over from points-type ignition systems to electronic ignition systems was because they were forced to change. As a result of stricter government-mandated emissions standards, lean air/fuel mixtures and mandatory use of exhaust gas recirculation (EGR) require an increase in spark-plug firing voltages, which are above the levels found in older points-type ignitions. However, many early electronic systems used mechanical controls, such as centrifugal and vacuum advance controls, for ignition timing. As technology advanced with time, these earlier versions were replaced by computer controls.

Along with producing stricter emission controls, automakers were also required to provide warranty coverage of 50,000 miles minimum for all emissions-related components (including ignition systems). Points-type ignition systems were not capable of lasting that long and had to be

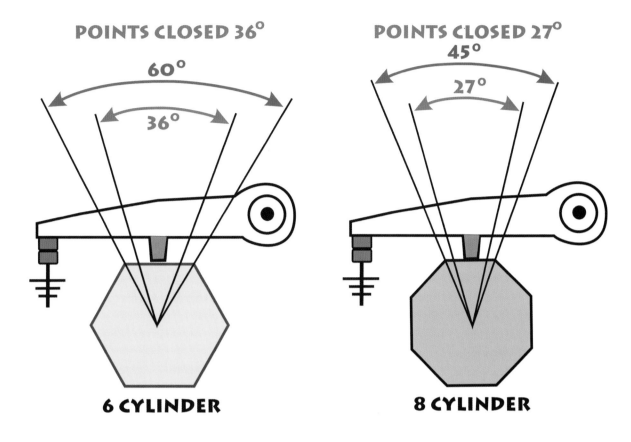

POINTS CLOSED 36°

60°

36°

6 CYLINDER

POINTS CLOSED 27°

45°

27°

8 CYLINDER

Fig 6-5. *Ignition dwell is longer on a six-cylinder engine than on a V-8. A dwell angle is the length of time the points spend closed, or charging, the ignition coil. The longer the dwell time, the higher the secondary voltage will be when the coil discharges across the spark plug.*

The test light: the universal tool for testing primary ignition coil switching. When connected across the primary circuit the flashing test light provides visual confirmation that the points, or ignition module, is turning the coil on and off.

IGNITION SYSTEMS

103

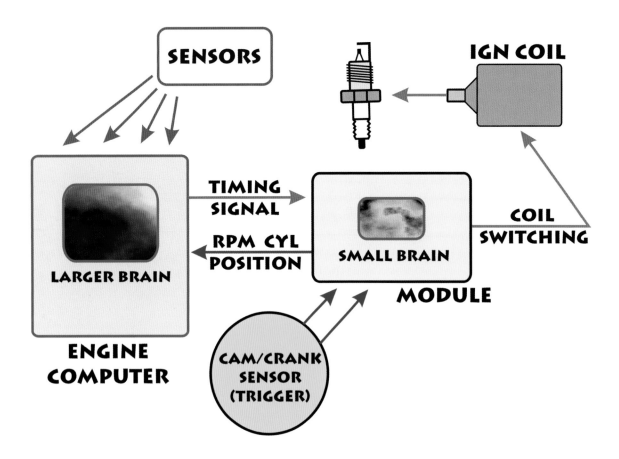

Fig 6-6. *The cam and/or crank sensors send information to the ignition module, which then passes it to the engine computer. The computer calculates the ideal spark timing signal and sends it back to the module, which then fires the coil.*

serviced every 10,000 miles or less. By contrast, since electronic ignition systems contained no moving parts, they were able to last 50,000 miles or more, meeting the government-mandated warranty period. In fact, maintenance intervals on late model coil-over-plug systems (no spark plug wires) oftentimes last until as many as 100,000 miles have been reached (and the maintenance usually amounts to nothing more than spark plug changes). By the early 1980s, most stand-alone electronic ignition systems were phased out and replaced with onboard computer-controlled ignition systems. The use of onboard computers eliminated the need for moving parts (the vacuum advance and centrifugal ignition advance mechanism) that controlled ignition timing in older electronic systems.

Ignition timing is now controlled electronically and calculated by input from engine computer sensors. Vehicles made after 1996 are equipped with OnBoard Diagnostics II (OBDII) programming that can detect ignition misfires, identify if a cylinder is having a problem (and if so, which one), and report the results of any malfunction as a trouble code. Most vehicles today have ignition modules that interface with engine management systems. A trigger signal sends information about engine rpm, cylinder identification, and crankshaft position to the module, which then processes the signal and sends it to the vehicle's computer. Then the computer determines the various parameters for ignition timing, including: engine speed, temperature, load, engine knock or ping, throttle position, intake air temperature, barometric pressure, air conditioning use, and transmission gear selection. The onboard computer crunches the numbers from the various digital inputs and sends a modified timing signal to the module, which then fires the ignition coil. Ignition timing control is quite sophisticated on late model vehicles; in fact, timing can be modified on individual cylinders by as little as 1/4 degree of crankshaft revolution.

TRIGGERS

For over 100 years, a set of points served as the "trigger" in conventional ignition systems. The points triggered the coil

The rotating magnet inside this distributor has eight points, or lobes, one for each cylinder. As it rotates past the pickup coil, voltage pulses are induced into the coil and sent to the ignition module.

on and off. Today, both electronic and computer-controlled ignitions use a transistor to switch an ignition coil on or off. There are only three types of triggers currently in use in newer model vehicles: an AC pickup coil, a hall-effect switch, and an optical sensor. The trigger is connected to the engine's crank or camshaft, where it produces a signal indicating engine speed which it sends to the ignition module. Some triggers are capable of providing crankshaft or camshaft position and cylinder identification as well.

Following is an explanation of how each of the various triggers operates and how to test their operation.

AC Pickup Coil

Whenever a wiring diagram shows an engine speed sensor with only two wires coming out of it, it's probably an AC pickup coil. It operates in the same manner as an alternator to produce AC voltage. A rotating series of magnets pass across a coil of wire inside the pickup coil. Each time one of the magnets passes the coil of wire, voltage is induced into the wire; as engine speed increases, the number of magnetic pulses per second increase. The ignition module or computer tracks the frequency of the pulses and, based on this information, calculates engine rpm (and in some cases, crank or camshaft position). As engine rpm increases, AC voltage also increases in addition to the increases in number of pulses. Although the increased AC voltage is

ignored by the computer, it's a good way to verify whether the AC pickup coil is working or not.

You can test AC pickup coils by using a voltmeter, ohmmeter, or logic probe. To test one using a voltmeter, disconnect the AC pickup coil and connect the meter leads to the pick-up wiring harness. Set the voltmeter to read AC voltage and crank the engine. As the engine turns, AC voltage should be produced—generally between 0.5 and 3 volts AC. A logic probe is just as easy and effective to use as a voltmeter. Connect the logic probe to the vehicle's battery. Leave the AC pickup coil connected and back probe one of the two wires coming from the pickup coil with the logic probe. Crank the engine and watch for a pulse on its LED. If no AC voltage is present, or if the logic probe doesn't show a pulse, you can always check a pickup coil's resistance using an ohmmeter. Resistance will vary depending on coil manufacturer, but a resistance reading somewhere in the 150- to 1,200-ohm range is generally okay. Always check for broken wires leading to the pickup coil or if there are metal filings on it, or problems with the air gap between the sensor and magnet, as well as a nonrotating distributor.

Hall-Effect Switch

One of the more common ignition triggers is a hall-effect switch, which consists of a sensor, magnet, rotating shutter

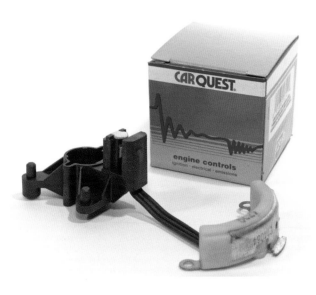

This hall-effect switch (left) is for a Ford vehicle. The white plastic connector (right) connects the switch to the ignition module. Courtesy of CARQUEST Auto Parts

Inside this optical distributor there are 360 holes on the outer ring of the metal disk—one for each crankshaft degree. The inner ring has four holes, one for each cylinder. The large hole is for cylinder number one in the engine's firing order.

blade, and three-wire connector (the latter distinguishes it from an AC pickup coil or other speed sensor on a wiring diagram). The three-wire connector has power, ground, and signal wires going to the computer or ignition module.

A hall-effect switch is powered by a reference voltage sent by the computer. It is always equipped with a magnet situated opposite the switch; between them is a series of rotating shutters (or blades)—one for each cylinder. As the blades rotate between the sensor and magnet, the magnet's magnetic field is interrupted and voltage drops at the sensor. The output signal from the hall-effect switch is a square wave, or series of on-and-off pulses. These pulses are sent to the vehicle's computer or ignition module, which use them to calculate engine rpm and crankshaft position.

To test a hall-effect switch, turn the ignition on and back probe each of its three wires with a voltmeter. A process of elimination will help identify what each wire is used for. The voltmeter display should indicate that one wire is reference voltage, which should have between 2.5 and 12 volts depending on year, make, and model of the vehicle being tested. Another wire is the ground wire and has no voltage; the third wire is a signal output wire from the hall-effect switch. The voltmeter will show either reference voltage or a value less than reference voltage, depending upon whether or not a shutter blade has stopped rotating between the hall-effect switch and the magnet. Turn the engine over by hand and watch the signal wire. If the hall switch is working, voltage will switch from reference voltage

to a lower voltage (sometimes 0 volts) and back. If a signal is not present, try the other wire that had voltage on it, since you may have read the reference wire by mistake.

In addition to testing a hall-effect switch with a voltmeter, a logic probe can also be used. Connect the logic probe to the vehicle's battery. Leave the hall-effect switch connected and back probe the signal wire. Crank the engine using the starter and watch the LED on the logic probe; it should indicate a pulse. If it doesn't, check to make sure there is both power ("reference voltage") and a ground source to the hall-effect switch, as well as for broken wires or loose connections.

Optical Sensors

An optical sensor is another form of trigger for an ignition module. It typically consists of an LED, a phototransistor, and a rotating metal disk with holes in it. (Some optical conversion kits that are used to replace a set of points have a plastic shutter instead of a metal disk.) These sensors provide trigger signals for ignition switching and generally can be distinguished from other speed sensors on a wiring diagram by the presence of four wires. The signal output from an optical sensor's computer takes the form of a square wave made up of on-and-off pulses (similar to a hall-effect switch).

When in use, a beam of light from the LED is projected through the holes in the disk onto the phototransistor. As the disk rotates, the spaces between the holes interrupt the light beam. Each time the light beam is interrupted, a pulse is

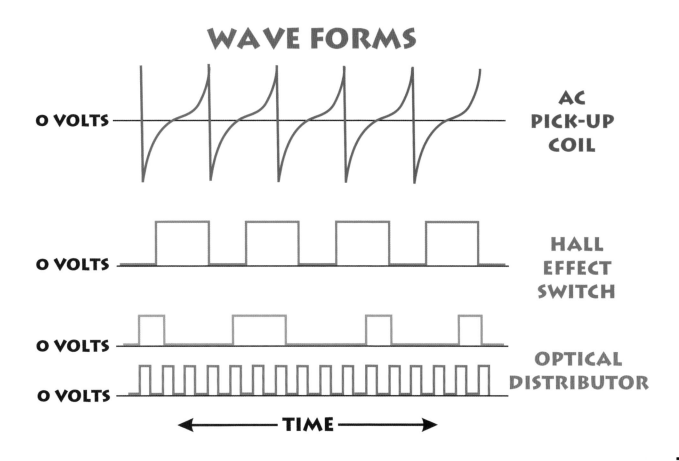

WAVE FORMS

AC PICK-UP COIL

HALL EFFECT SWITCH

OPTICAL DISTRIBUTOR

0 VOLTS

0 VOLTS

0 VOLTS

0 VOLTS

◄── TIME ──►

Fig 6-7. *An AC pickup coil's wave form shows AC voltage transitioning between negative and positive as the sensor is rotated. Both the hall-effect switch and optical sensor produce a square wave. The optical sensor has two signals—cylinder identification and engine rpm (lower wave form).*

generated by the sensor's processor. Some optical sensors are really two sensors in one: one measures crankshaft angle, the other measures camshaft position and identifies which cylinder is the number one cylinder in the engine's firing order.

To test an optical sensor, use a voltmeter to measure voltage on all the wires going to the sensor. With the ignition key turned to ON, one wire should read 12 volts and the ground wire should read 0 volts. However, it's difficult to determine if both signal wires are working using only a voltmeter. A logic probe is the better tool for detecting the presence of pulses on each signal wire. With the logic probe connected to the battery, back probe each signal wire and crank the engine. There should be a pulse on each wire; if there is no pulse, check for oil or dirt blocking the holes on the metal plate. Also, make sure the sensor has both power and ground.

IGNITION MODULES

In its most simplistic form, an electronic ignition module is nothing more than a modern replacement for a set of

points. Points receive engine rotational speed information by direct mechanical means—they open and close, turning the ignition coil on and off. An ignition module performs the exact same function, only it uses a trigger signal. Once a trigger signal (or wave form) is produced by the engine's speed sensor, the ignition module processes it and then fires the coil. Later model vehicles use onboard computers in conjunction with ignition modules to control coil switching. Either way, the basic operation of an ignition module is equivalent to a set of points.

GENERIC IGNITION MODULE TESTING

When electronic ignitions were first introduced, many technicians had trouble diagnosing no-spark problems. Without any moving parts to check, it was visually impossible to determine if primary coil switching was occurring. Fortunately, today, even though you can't see a transistor operating, you can see the results of primary switching—just like on a set of points. The first step in the process of determining if an ignition module is operational is a testing procedure originally used on

107

While these ignition modules look different, they all perform the same function—switching the coil on and off. Vehicles built after 1984 may also use onboard computers that work in conjunction with ignition modules to perform coil switching.

conventional ignition systems. Just loop a test light across the ignition coil's primary wires and crank the engine. A flashing/flickering test light tells you two things—(1) the ignition module is switching the coil on and off; and (2) the trigger is sending a signal to the module. This test works regardless of the type of ignition module being used, whether DIS or a coil-over-plug system (see Figure 6-8).

However, if the test light doesn't flash (because no primary switching is occurring), the ignition module should be checked for power, ground, and trigger input(s). Because each year, make, and model of vehicle is different, you must consult a wiring diagram to determine module inputs/outputs and wire colors. Powers and grounds can be checked using a digital voltmeter. Make sure ground voltage is close to 0 volts and the power wires have battery voltage. Check for power and grounds with the key turned to ON. Then be sure to recheck them with the engine cranking (especially the grounds), since ground voltages that checked out okay with the key turned to ON may increase once the starter is rotating. A trigger signal will either come directly

from the trigger (hall-effect switch, AC pickup coil, or optical distributor) or from a computer on some vehicles. If any of the module's wires don't have correct electrical values for voltage, ground, or trigger signals, they must be repaired before continuing further testing. However, if all the wires going to the ignition module have correct values, the module needs to be replaced.

IGNITION MODULE "TAP" TESTING

Another useful method for testing an ignition module is to provide a substitute trigger signal to determine if the coil can produce a spark. This method is called a "tap" test, since a grounded (or powered) test light is used in place of a trigger to "tap" on the trigger inputs to the ignition module. If the module is okay, and all the power and ground wires going to it are also good, it should fire the ignition coil. Ignition systems equipped with single coils are typically easier to perform tap tests on, but a tap test will even work on DIS and coil-over-plug ignitions. (Because these systems use cam and crank position trigger signals, the procedure

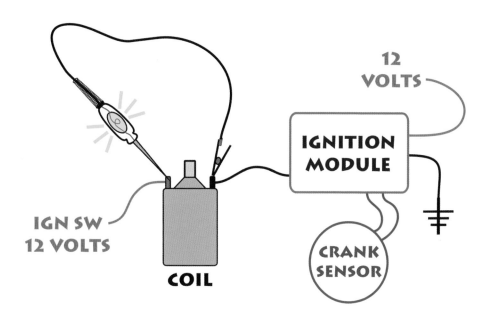

Fig 6-8. *The test light provides visual confirmation that the module is sending a primary switching signal to the ignition coil. This test will work on both single coil and DIS ignition systems.*

IGN SW
12 VOLTS

COIL

12 VOLTS

IGNITION MODULE

CRANK SENSOR

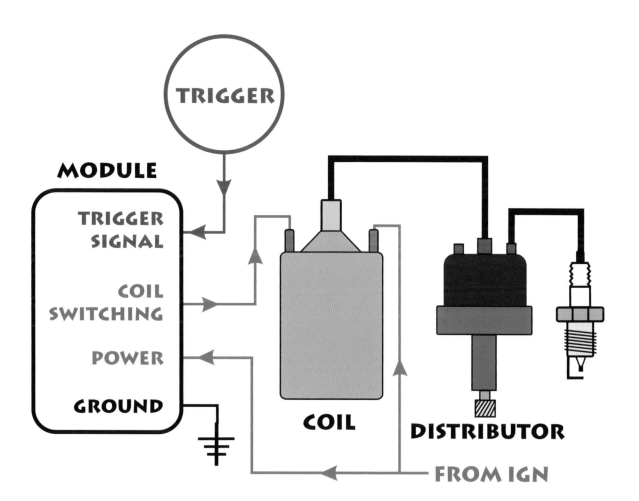

TRIGGER

MODULE

TRIGGER SIGNAL

COIL SWITCHING

POWER

GROUND

COIL

DISTRIBUTOR

FROM IGN

Fig 6-9. *A generic ignition module requires the presence of four things in order to fire a coil: power, ground, trigger input, and coil switching. Wiring diagrams identify what each of the wires at the ignition module is used for.*

may be more complex to perform, often requiring two test lights and a timed tapping sequence.) If you don't get a spark after performing a tap test, always check for the presence of power and a good ground or bad ignition coil. Also, make sure the test light used to perform the tap test has a resistance value less than 10 ohms, since test lights with bulbs having more resistance may not pass enough current to make a tap test work. (See Chapter 3 on electronic testing tools.)

Following are some of the more popular ignition tap tests relating to specific manufacturers.

General Motors

The following tap test works on most 1981 to 1993 GM vehicles equipped with a distributor. On these vehicles, the ignition module is located inside the distributor and may have four, five, or seven terminals. All modules have green and white wires connecting them to the AC pickup coil. Connect a spark tester (or old spark plug) to the coil to check for spark output. Then unplug the pickup coil wires at the module (leaving all other wires connected) and turn the ignition key to the RUN position. Using a hot test light (alligator clip connected to battery positive), tap back and forth on the module terminals where the green and white pickup coil wires were originally connected. If the module is good, the coil should produce a spark. If the tap test is successful, the pickup coil is probably bad, but don't forget to check for a bad coil before replacing parts. GM pickup coils should have between 500 and 1,500 ohms of resistance. Occasionally, the fine wires inside the pickup coil intermittently short, or open, causing ignition misfire. When checking a pickup coil with an ohmmeter, try hitting the pickup coil with a screwdriver while watching the meter reading—if it changes, buy a new pickup coil.

There is a different tap test for 1987 to 1993 2.8- or 3.1-liter GM vehicles with DIS ignitions. The three ignition coils that make up the coil pack are connected directly to the module. Unplug the three-wire connector on the module from the crank sensor and remove the spark-plug wires from the coil pack. Then connect a jumper wire from any terminal of one coil to ½ inch below the opposite terminal of that same coil—this is where the spark will jump. Be sure to use three jumpers (one for each coil), since any of the three coils may spark when this tap test is performed. Turn the ignition key to ON and, using a hot test light, tap module terminals A and C (where the crank sensor plugged in). If the module is good, you'll get a spark that will jump between any of the three coil pairs.

If you get a spark, check the resistance of the crank sensor at the module connector—it should be between 900

To perform a tap test on an ignition with an AC pickup coil, leave all connectors plugged in except the pickup coil. Using a hot test light, tap back and forth between the module terminals where the pickup coil was originally plugged in. If the module is good, the coil should spark.

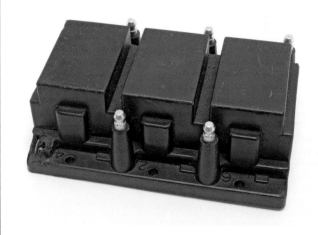

This GM coil pack has three coil pairs for a V-6 engine. The primary wires are underneath and connect directly to the ignition module.

and 1,200 ohms. (You can also measure the sensor's AC voltage output on the purple and yellow wires at the module. With the engine cranking, AC voltage should be between 0.3 to 1.5 volts.) If you don't get a spark, check for power and ground on the two-wire connector at the module. In addition, check the resistance on each ignition coil. Primary circuit resistance should be between 0.3 and 2.0 ohms; secondary resistance should be between 5,000 and 10,000 ohms. If you have power and a good ground at

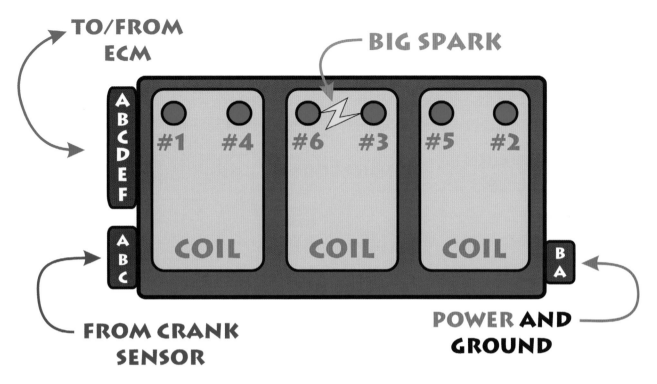

TO/FROM ECM

BIG SPARK

A B C D E F

A B C

#1 #4   #6 #3   #5 #2

COIL   COIL   COIL

B A

FROM CRANK SENSOR

POWER AND GROUND

Fig 6-10. *This module and coils are used on 2.8- and 3.1-liter GM cars. With everything plugged in, tapping on terminals A and C at the three-wire connector should produce a spark from one of the coils if the module is good.*

the module and the coil pack has correct resistance, the module is most likely bad. This test can also be performed with the module/coil pack out of the vehicle. Just be sure to supply power and ground to the two-wire connector at the module, and then perform the tap test as described, as in Figure 6-10.

Ford Motor Company

There is a different tap test for 1975 to 1990 Ford Dura Spark ignition systems. The Dura Spark system is best identified by what's inside the distributor. Look for an AC pickup coil, mechanical advance (weights and springs), and a vacuum-advance. Dura Spark modules are typically mounted inside the engine compartment, usually on the fender well.

To test the module, disconnect the ignition coil's high-voltage lead to the distributor and install a spark tester in its place. With all wires connected to the module, back probe the orange wire (module side of the wiring harness) with a test light. Crank the engine over and leave the ignition key in the RUN position—this initializes the module. (Don't use a remote starter to crank the engine or this tap test won't work.) Next, tap the test light to the positive or negative battery terminals. If the module is good, the ignition coil should produce a spark. If you get a spark, there are three possibilities: (1) the AC pickup coil is shorted or open, (2)

the wires from the distributor to the module are not connected, or (3) the distributor is not turning.

If you don't get a spark, all the wires going to the module must be checked with the engine cranking. With both the two-wire and four-wire connectors plugged into the module, back probe each wire. The black wire should read 0 volts, the purple and orange wires should have about 0.5 volt AC, and the green wire should produce a pulsing signal that can be checked with either a logic probe or hot test light. The two-wire connector should have battery voltage in both the RUN and START ignition positions. If the purple and orange wires don't have 0.5 volt AC with the engine cranking, the AC pickup coil may be bad. The pickup coil's resistance should read between 400 and 1,300 ohms. The ignition coil's resistance should also be checked; primary resistance should be 0.8 to 1.6 ohms, while secondary resistance should read between 7,700 and 10,500 ohms (see Figure 6-11).

Ford used a thick film ignition (TFI) from 1982 to 1993. TFI systems use a spark output (SPOUT) connector. This is a removable connector that disconnects the ignition module from the vehicle's computer. The first step in diagnosing a TFI no-spark problem is to unplug the SPOUT connector. Then crank the engine and check for a spark. If you get a spark, check for a burned fusible link, a

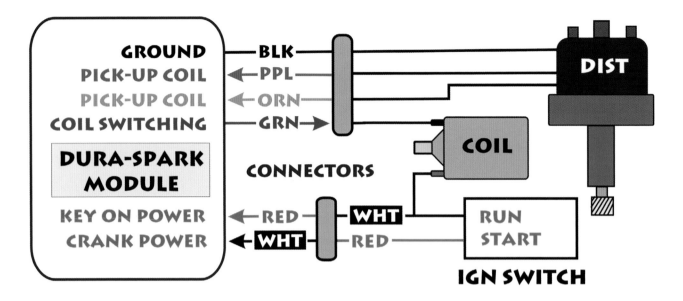

Fig 6-11. *If this Dura-spark module is good, tapping on the orange wire will produce a spark at the ignition coil. However, before tapping, be sure to initialize the module by cranking the engine over using the ignition key; otherwise the tap test won't work.*

bad ground to the ECM, or a bad ECM. If there is no spark, leave the SPOUT disconnected and then crank the engine. (Don't use a remote starter or the test won't work.) Cranking the engine with the key will initialize the module. Using a test light, back probe the wire labeled 1 (closest to the distributor cap) at the module as in Figure 6-12. Tap the pointy end of the test light to the battery's negative or positive terminal—a spark from the ignition coil should occur and the fuel pump should run.

If there is a spark and the distributor turns, the ignition module is good but the hall-effect switch is not. If there is no spark and the ignition coil tests okay (see the section on generic coil testing discussed earlier in this chapter), the wires at the module need to be checked for correct values (see Figure 6-12). Module terminals 1 and 2 should each have about 5 or 6 volts with the engine cranking. Terminal 3 should read about 11 volts with the engine cranking, and terminal 4 should read about the same with the key in the RUN position. Terminal 5 should have a pulsing signal that can be checked with a logic probe. The last terminal, labeled 6, should read 0 volts, as it is the ground. If all the wires connected to the module have the right readings, it's time to get a new module.

Chrysler

The next tap test is for Chrysler vehicles equipped with carburetors and distributors that were manufactured between 1972 and 1989. The distributors on these vehicles

Shutter blades (top right) inside the distributor pass between the hall-effect switch (on the left) and a magnet. The hall-effect switch is connected (with a brown connector) to the Thick Film Ignition (TFI) module that is mounted on the side of the distributor.

used either a single or dual AC pickup coil; there are tap tests for both types. In general, vehicles with two-barrel carburetors usually have a single AC pickup coil, while those with four barrels have dual pickup coils. If dual pickup coils are in used, one is the RUN pickup and the other is the START pickup. Either an ignition module or a spark computer may be used to fire the coil. A single or dual

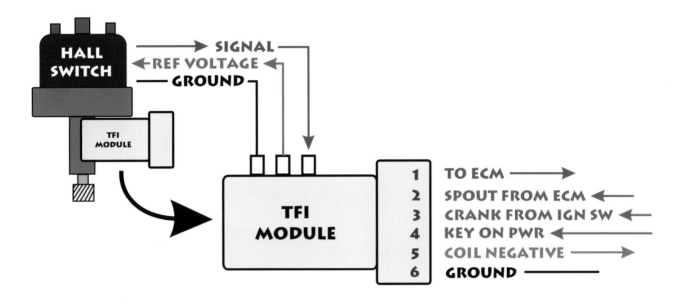

Fig 6-12. *Tapping on terminal 1 of TFI ignition module (the wire that goes to the ECM) at the module will produce a spark at the ignition coil if the module is good. Don't forget to unplug the spark output (SPOUT) connector before doing this tap test.*

This Chrysler ignition module uses a large transistor mounted in a gold-colored aluminum heat sink (upper left of module). Courtesy of CARQUEST Auto Parts

ballast resistor (like those used in points-type ignitions) may be used in series with the ignition coil. To perform a tap test on a single AC pickup coil distributor, turn the ignition key to the RUN position. Using a hot test light, tap back and forth between the pickup coil wires at the distributor connector. Watch for a spark from the coil.

You can check the RUN pickup coil on a dual pickup coil distributor in the same manner as a single pickup coil distributor. To check a START pickup coil on a dual coil system, the engine must be cranking when the tap test is performed. The large connector at the distributor connects to the START pickup coil; the smaller one connects to the RUN coil. Only one pickup coil is used at a time—either the RUN coil (engine running) or the START coil (engine cranking).

There is an air gap adjustment that must be made on both single and dual pickup coils. The single pickup coil air gap should measure 0.008 inch for 1972 to 1976 vehicles and 0.006 inch for 1977 to 1986 vehicles; the dual pickup coil air gap should be 0.008 inch for all years. Pickup coil resistance should be between 50 and 900 ohms. The single ballast ignition resistor's resistance should measure 1.25 ohms, while the dual ballast resistor's resistance should be 5 ohms (run side) or 0.5 ohm on resistors with exposed windings, and 1.25 ohms for resistors with sealed windings. If after performing the tap test there is no spark, check the wires at the module or computer for power and grounds.

Chrysler vehicles equipped with fuel injection systems manufactured between 1984 and 1995 use a hall-effect switch inside the distributor. In addition to the hall-effect switch, 1984 to 1987 models have two onboard computers—a logic module (located in the passenger-side kick panel inside the car) and a power module (located next to the battery). Later model vehicles produced from 1995 to 1998 use a single computer called a single board engine controller (SBEC). All of these vehicles use an auto shut

An auto shut down (ASD) relay is hidden inside the power module (top). In addition to controlling the fuel injectors and charging system, this power module serves as the ignition module for early fuel-injected Chryslers. The brains of the engine management system is the logic module (bottom). Courtesy of CARQUEST Auto Parts

down (ASD) relay to power the fuel injectors, fuel pump, and ignition coil. The ASD relay is controlled by the onboard computer and is turned on for only two seconds when the key is in the RUN position. If the key is cycled between RUN and START positions three or more times (usually because the car won't start), the computer won't turn on the ASD relay until it receives a crank signal from the distributor. Because of this potential problem, you need to bypass the computer's control of the ASD relay before performing a tap test on these vehicles.

To do this, the relay's control wire needs to be grounded. The ASD relay can be either internal or external to the computer, but you must consult a wiring diagram to locate the wire that triggers the relay. The relay control wire to the computer will be labeled "ASD control" or "ASD relay." Once you locate the control wire, use a grounded test light to trigger the ASD relay. With the ASD relay energized, check for 12 volts at the coil and listen for the fuel pump to run. With the ASD relay on, the tap test can now be performed.

There is a separate ignition tap test for Chrysler vehicles produced between 1984 and 1987 with 2.2- and 2.5-liter engines. These vehicles use a hall-effect switch inside the distributor and a three-wire connector to the onboard computer.

In addition, two computers are used—a logic module and a power module. The ASD relay is inside the power module; the relay's trigger wire is blue with a yellow stripe. To perform a tap test, unplug the three-wire connector at the distributor and turn the ignition key to ON. Using a grounded test light, tap on the gray wire going to the logic module. This should produce a spark at the coil. If you get a spark, plug the distributor back in. With the engine cranking, use a logic probe to check for a signal from the hall-effect switch (gray wire). If no switching signal is present, check each of the three wires going to the distributor with the connector unplugged. The orange wire should have 8 volts, the gray wire 5 volts, and 0 volts on the black wire (ground). If the wires all check out okay, it's time to buy a new hall switch. If there is no spark, the power and ground wires at the power module all need to be checked for correct values. Use a wiring diagram specific to the vehicle being worked on in order to identify wire colors and corresponding electrical values. If the wires are OK, the power module may be bad.

The tap test for Chrysler vehicles made from 1988 to 1995 is basically the same as the previous test. These vehicles use a single computer Chrysler called a single board engine controller (SBEC). Before performing this tap test,

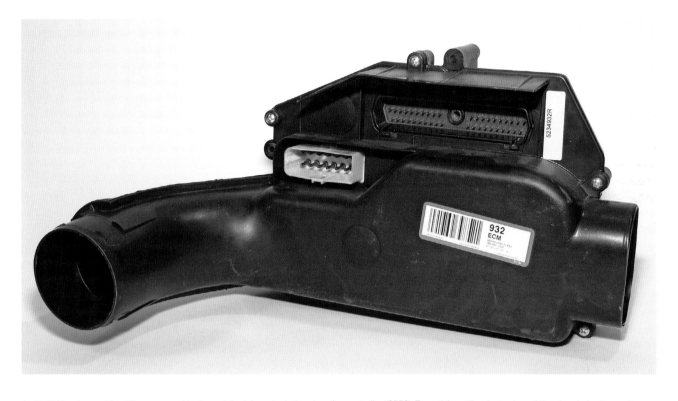

In 1988 Chrysler combined the power and logic modules into a single board engine controller (SBEC). To cool down the electronics, all the air entering the engine must pass through air ducts at each end of the SBEC. Courtesy of CARQUEST Auto Parts

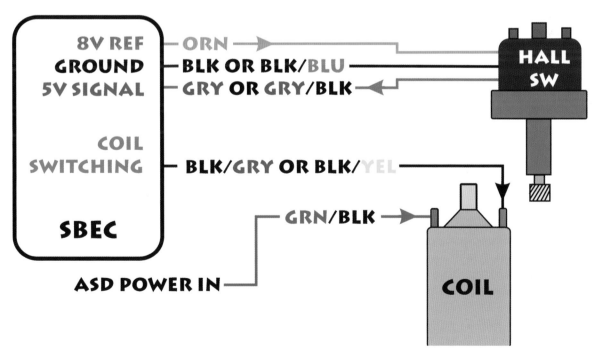

Fig 6-13. After grounding the ASD relay on this Chrysler ignition system, check for power at the coil before doing this tap test. If the relay is not working, the tap test won't produce a spark.

These replacement ignition wires are constructed using high-grade silicone that can withstand 500 degrees Fahrenheit, or under-hood temperatures. They are available in either a carbon impregnated fiberglass core for radio frequency interference (RFI) suppression or a plated copper core for racing. Courtesy of Mallory Ignition and Summit Racing Equipment

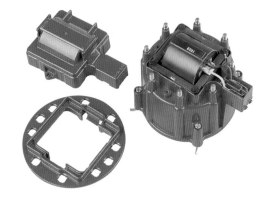

This high energy ignition (HEI) upgrade kit for GM vehicles uses an alkyd distributor cap with brass terminals. These replacement parts exceed OEM specifications. Courtesy of Summit Racing Equipment

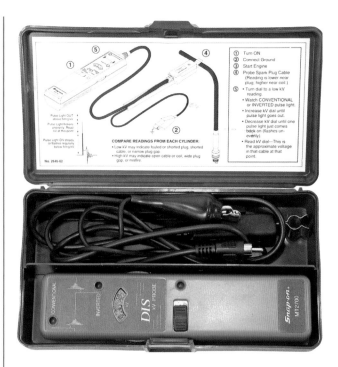

This Snap-On MT2500 kV spark tester can be used on both conventional and DIS ignition systems. Flashing LEDs and a rotary dial indicate spark voltage levels.

the ASD relay must be triggered by grounding the control wire with a grounded test light; the ASD relay is always external to the computer in these model years. To perform a tap test on these vehicles, unplug the three-wire connector at the distributor. With the ignition key in the RUN position, use a grounded test light to tap on either the gray or gray/black wire (at the hall-effect switch) on the harness side of the connector. The coil should produce a spark. If you get a spark, plug the distributor back in and use a logic probe to check for a signal from the hall-effect switch while the engine is cranking. If no switching signal is present, check each of the three wires going to the distributor with the connector unplugged. The orange wire should have 8 volts, the gray or gray/black wire should have 5 volts, and there should be 0 volts on the black or black/blue wire. If all the wires check out okay, you need to buy a new hall-

effect switch. If you didn't get a spark, the power and ground wires at the power module all need to be checked for correct values. Use a wiring diagram specific to the vehicle you are working on to identify wire colors and corresponding electrical values. If the wires are OK, power module may be bad.

SECONDARY IGNITION CIRCUIT

So far we've dealt only with primary ignition switching; now the focus shifts to the secondary ignition circuit. When a spark is produced at the ignition coil, it must go from there to the proper spark plug in the engine's firing order. On vehicles equipped with distributors, the spark travels through the coil's high-voltage wire to the distributor cap and then to the rotor. The rotor directs the spark to the terminal inside the distributor cap that corresponds with the cylinder requiring a spark, based on the engine's firing order. The configuration of crankshaft and camshaft determines the ignition firing order of the engine's cylinders. As each cylinder reaches the top of its compression stroke, its air/fuel mixture is ready to be ignited by the ignition system. The spark from the ignition coil necessary for combustion jumps the air gap between the rotor tip and cap terminal, and then travels along the spark plug wire to the spark plug

IGNITION SYSTEMS

at each cylinder. On vehicles without distributors, the correct firing order is programmed into the ignition module or onboard computer. The coil pairs on a DIS system fire in a sequence directly matching the firing order of the engine. In coil-over-plug systems, individual coils are automatically triggered in correct firing order by the vehicle's computer.

Voltage traveling to the spark plugs must overcome the high resistance of the plug's air gap between the center and ground electrodes. When the engine's cylinder is on its compression stroke, the resistance of the air gap is increased; consequently, even more voltage is required for the spark to jump the gap. Older cars with points-type ignition systems may require as much as 15,000 to 20,000 volts to fire a spark plug with the throttle open. Vehicles designed with three-way catalytic converters have extremely lean fuel mixtures that are hard to ignite. In addition,

exhaust gas recirculation (EGR) is added to the combustion process, causing spark-plug firing voltage to be even higher. This combustion chamber environment increases the spark plug's air gap resistance well beyond what older ignition systems were able to provide. In fact, actual firing voltage may reach 50,000 or more on some late model systems.

Because of high ignition voltage requirements, secondary voltage tries to make its way to battery ground via the path of least resistance. The secondary wires, distributor cap, and rotor all eventually wear out with age and must be inspected periodically for damage and/or arcing caused by voltage leaks. Check the inside of the distributor cap for carbon tracking, which appears as shiny, black lines. Also, green- or white-colored corrosion can form on the rotor tip and cap terminals. Secondary wires should be visually inspected for cracks, brittle insulation, or loose connections at the distributor cap or spark plugs. The

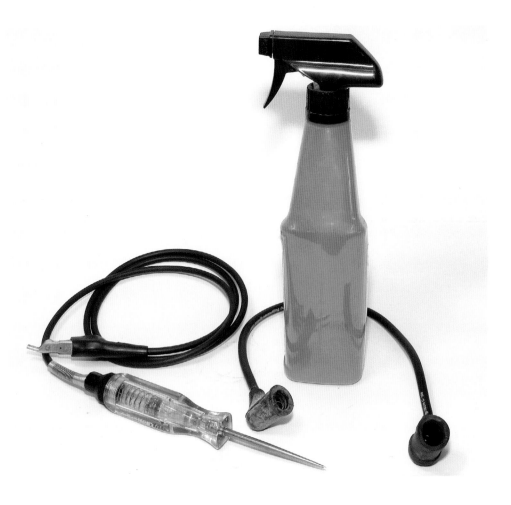

After spraying secondary ignition wires with water, a grounded test light can be used to find bad wires or spark plug boots. When placed close to the leaky wire, the tip of the test light will attract the high-voltage spark causing a misfire.

cap and rotor can be checked using an ohmmeter, but again, this will only confirm if the connections are open. Remember, they may check out all right with an ohmmeter, but could break down under load, causing misfires. High voltage ignition cables can also be checked using an ohmmeter, but resistance values vary widely between manufacturers, so be sure to consult a service manual for specific ohms per inch/foot specifications. Keep in mind that using an ohmmeter for resistance checking has limited value. When a driver opens the engine's throttle, engine compression increases to maximum. The added resistance may cause a bad wire to leak voltage to ground without going to the spark plug first.

There are several low-dollar methods for diagnosing secondary ignition misfires. An inductive timing light clamped around an ignition coil's high-voltage wire, which goes to the distributor cap, can serve as a poor man's ignition scope. To check for ignition misfires, connect the timing light to the coil wire (between the coil and distributor cap) and start the engine. Point the timing light at your face and with the engine running, snap the throttle open. If the flashing light skips a few flashes, either a bad coil wire or a primary switching problem may be the cause of the misfire. If no misfire is detected with a timing light, but the engine is obviously misfiring, move the inductive clamp of the timing light to each individual ignition wire and repeat the test. This will help isolate the bad spark plug wire. This test will also work on DIS ignition systems—just be sure to clamp the timing light over each pair of plug wires until the ignition coil that is causing the misfire is found. On coil-over-plug systems, the coil must be removed and a temporary plug wire must be installed between the coil and plug in order to provide a place to connect the timing light for the test.

Another method for finding bad ignition wires requires only a grounded test light and some water. Using a spray bottle filled with water, get the ignition wires really wet. The water helps magnify a spark-plug boot or wire that is leaking voltage. Connect a test light to ground and start the engine. Move the tip of the test light along all the ignition wires while listening to the engine. When the test light is placed close to a bad section of wire, secondary voltage will jump to the sharp tip of the test light, causing the engine to misfire. If it's dark outside when performing this test, you can see, as well as hear, where the problem is occurring.

CHAPTER 7
FUEL INJECTION SYSTEMS

With a final gasp of breath, the use of carburetors in U.S. production vehicles died in 1994. The widespread use of electronic ignitions beginning in 1975 and EFI systems in 1984 was driven by ever-tightening emissions standards. Although there was a brief period in the early to mid-1980s when manufacturers used electronically controlled carburetors to meet emissions requirements, these systems were fraught with problems, both from poor design and lack of training for technicians working on them. Fortunately, these early production attempts to control fuel delivery are a thing of the past. Nowadays, all vehicles made in or for the United States have EFI systems.

In addition to cars and light trucks, many nonautomotive applications are currently experiencing a transition phase from carburetion to EFI, including motorcycles, scooters, and lawn tractors. Today, a technician or home mechanic who is without an understanding of how EFI

This ECM is out of an early 1980s GM vehicle. The combination of government-mandated emissions control and the availability of cheap computers allowed auto manufacturers to switch from carburetors to electronic fuel injection systems.

The computer-controlled carburetor was an intermediate step between the old carbureted and the new electronic fuel injection technology.

works has severely limited ability to perform even basic electrical and electronics diagnostics.

However, working on an EFI system doesn't have to be as complex as many people believe it is. Just like electronic ignition systems, different EFI designs share many similarities, so once a technician understands the basics of EFI operation, most types should be relatively easy to diagnose.

CARBURETORS

All gasoline-powered engines need only two basic ingredients to run—the correct amount of fuel for any given rpm and throttle opening, and a spark from the ignition coil at the right time. A carburetor's fuel-delivery system is made up of separate fuel circuits, each with a specific job to perform. During startup of a cold engine, the choke circuit adds extra fuel and air to keep the engine from stalling. Once the engine reaches operating temperature, the choke circuit shuts down. At idle (as the throttle is opened), the idle mixture from the screw and transfer port circuits provides additional fuel as required by the engine. As airflow into the engine increases, the main fuel circuit regulates fuel flow in relation to the amount of throttle opening. If the throttle is opened suddenly, an accelerator pump circuit squirts fuel directly into the intake manifold. Without an accelerator pump, the faster moving air would get to the intake valve and cylinder before the fuel, causing a flat spot and backfire during acceleration (air is about 400 times lighter than gasoline).

Although carburetors have provided fuel delivery services for well over 100 years and have always basically functioned pretty well, there are some things they simply can't handle. A carburetor is basically a hunk of aluminum with a bunch of holes drilled into it. As air pressure within the carburetor changes, fuel and air flow through the various holes and into the engine. However, a carburetor's ability to deal with constantly changing operating environments is limited with regard to reactions to changes in altitude and compensation for engine temperature. They also lack precise fuel control for emissions purposes and create overall excessive fuel consumption during steady-state engine operation and acceleration. In a word, carburetors are just too dumb to continue being useful for providing accurate air/fuel mixtures in modern vehicles. What is needed is a fuel delivery system with some brains.

ELECTRONIC FUEL INJECTION

In addition to performing all the functions of a carburetor, the EFI system also controls engine idle speed and various ignition system timing functions. It regulates fuel delivery using electromagnetic valves (fuel injectors) that open electronically for varying lengths of time. When the fuel injectors are turned on, they spray fuel into the engine. The amount of on time is called injector pulse width, and the longer this is, the greater the amount of fuel injected into the engine. While all EFI systems use fuel injectors that operate in the same manner, there are different EFI designs and computer strategies in use today.

Throttle body injected (TBI) engines use only one or two injectors located where the carburetor used to be. After fuel is injected into the throttle body, it travels through an intake manifold before reaching each cylinder. Because the air/fuel mixture has to travel varying distances to reach the engine's individual cylinders, this system has inherent fuel distribution problems.

By contrast, a port fuel injection (PFI) engine uses an individual injector for each cylinder, making it more efficient

FUEL INJECTION SYSTEMS

This Ford throttle body and injectors make up a modern EFI system, providing better performance and fuel mileage than the carburetors of the past.

Some Bosch fuel injection systems use a combination of mechanical and electronic components in their systems. Pictured is a mass airflow (MAF) sensor and fuel distributor for a K Jetronic system.

than a TBI system. The PFI system sprays fuel directly at the back of the engine's intake valves; consequently, fuel delivery is more precise because it has less distance to travel before reaching the combustion chamber. Eventually, PFI systems were replaced with sequential port fuel injection (SPFI) systems. Each cylinder still has an individual injector, however, the injectors on these systems are pulsed in accordance with the engine's sequential firing order. These systems provide more accurate fuel delivery, which in turn helps reduce exhaust emissions and increase fuel mileage.

Finally, one other flavor of fuel injection systems exists—a Bosch KE Jetronic. This system uses an airflow sensor with a metal plate that moves in relation to engine airflow. As the throttle is opened, the sensor measures the increased air entering the engine. A mechanical fuel distributor then delivers fuel to each injector. The injectors are simple pressure relief valves and use no electronics. These systems are found only in European cars.

TAKING EFI FOR A TEST DRIVE

For an electronic control module (ECM) to control injectors and ignition timing, it must receive information about engine rpm, engine temperature, altitude, air temperature, driver demand, and other inputs. This information is passed along to the ECM via its sensors, the ECM uses this information for overall engine management.

How a fuel injection system operates is worth a closer look. Taking a quick virtual drive in a modern fuel-injected car will illustrate how an EFI system operates.

Starting

Imagine yourself on a nice fall day sitting in your new car, which, naturally, is equipped with a modern EFI system. When the ignition is first turned on but before the engine is started, the ECM checks in with various sensors to determine engine coolant and air temperature, as well as barometric pressure. The sensors report a cool 50 degrees Fahrenheit outside ambient temperature. Because the car has been sitting overnight, engine coolant temperature is the same as the air temperature. Since your imaginary car was parked near the beach, the barometric pressure (BARO) sensor indicates to the ECM that barometric pressure is at sea level. The ECM turns on when the key is turned to the RUN position. Because engine temperature is cold and air pressure is at sea level, the ECM adjusts the injector's initial pulse width in accordance with internal programming to ensure a rich fuel mixture for cold starting. The ECM also opens an idle air controller (IAC) valve to allow extra air to enter the engine, causing a fast idle. This keeps the engine from stalling until it warms up. When the ignition key is

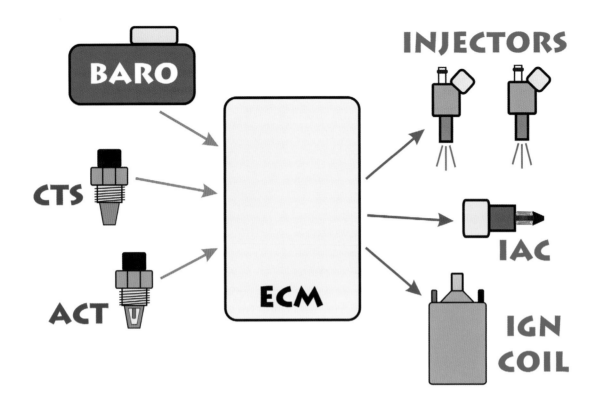

Fig 7-1. *Just like a fast idle cam and choke circuit on a carburetor, a coolant sensor, air temp sensor, BARO sendor, ECM, and injectors serve to provide a rich fuel mixture for cold starting. The idle air controller (IAC) increases engine idle speed during cold startup so the engine won't stall.*

turned to the START position, the ECM gets a signal from the crankshaft and camshaft position sensors as soon as the engine begins rotating. The crankshaft sensor sends an engine rpm signal and the camshaft sensor identifies which cylinder is next in the firing order. During cold startup, the ECM pulses (turns on) all of the injectors during every other crankshaft revolution. However, once the engine warms up, the injectors are pulsed in the same sequence as cylinder firing order—just like the spark plugs. Each injector is pulsed just before the intake valve opens for that cylinder. This fuel delivery strategy, called SPFI, decreases exhaust emissions while increasing horsepower, thereby helping out in the fuel economy department.

Warming Up and Driving

Once the engine warms up, the fuel mixture must be leaned out so there is less fuel and more air. The ECM does this by turning the injectors on (reducing injector pulse width) for a shorter amount of time. The ECM also lowers engine idle speed by closing the IAC valve. However, once the automatic transmission is put into DRIVE, the ECM immediately changes ignition timing and slightly opens the

IAC to maintain engine idle speed. This action provides a seamless idle speed to the driver when shifting from PARK to DRIVE. When the accelerator is pressed, the ECM checks several inputs to determine injector pulse width and ignition timing.

The TPS senses how far and how fast the throttle opens. The MAF sensor calculates engine load by measuring the amount of air entering the engine. MAP sensors measure engine vacuum, another indication of engine load. The ECM uses TPS, MAF, and MAP sensor inputs, as well as engine rpm, to determine injector pulse width and ignition timing. The ECM calculates all inputs by consulting its computerized internal dictionary, or "look-up tables," which contain information about how long to keep the injectors on for specific driving conditions. For example, if the TPS opens at a moderate rate, and engine rpm are coming up slowly, the ECM increases the injector on time gradually. However, if the throttle is opened suddenly, the TPS signal voltage goes up instantly and the injectors are given a long on-time (but only briefly!), thereby providing the same shot of extra fuel as a carburetor's accelerator pump.

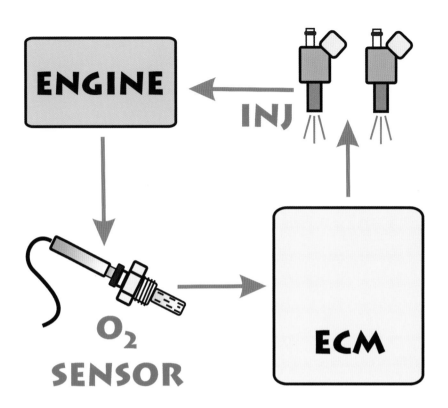

Fig 7-2. *The informational loop—consisting of inject, sense, and control—works to maintain the correct air/fuel ratio required by a vehicle's catalytic converter. Since the loop is endless while driving, it's often referred to as a closed loop.*

Normal Operating Temperature and Driving

With the engine at normal operating temperature and a steady throttle, the ECM goes into what is known as closed loop operation. The interaction between ECM, fuel injectors, and oxygen (O_2) sensors is often called a feedback loop or closed loop control. During closed loop operations, the ECM pulses an electronic signal to the injectors to add specific amounts of fuel in order to lower harmful exhaust gases. As fuel is burned along with air in the engine's combustion chamber, the resulting exhaust gases travel through the exhaust system where the unburned exhaust gas oxygen is measured by the O_2 sensor. The O_2 sensor sends a signal voltage to the ECM, which then modifies the injector pulse width to maintain a 14.7:1 air/fuel ratio. This air/fuel ratio is ideal for keeping exhaust emissions low. The sequence of inject, combust, sense, and control creates an informational feedback loop. This operating mode is also refered to as closed loop or closed loop control.

The ECM can even adjust fuel delivery based upon humidity, outside air temperature, and altitude. The MAF sensor discerns both the temperature of the incoming air and, indirectly, the amount of moisture (humidity) it contains. If you decide to head into the mountains for the day, the barometric pressure sensor changes its signal sent to the ECM. As you go higher in altitude, air density decreases and the ECM leans out the air/fuel mixture by decreasing the injectors' on-

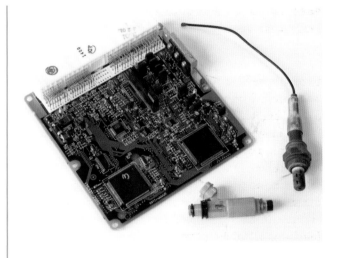

Three components—an ECM, injector, and oxygen (O_2) sensor—compose a closed loop fuel-control system. A computer and fuel injector have taken the place of older, electronically controlled carburetors.

time, so the amount of fuel the engine uses matches the reduced oxygen found at higher-than-sea-level altitudes.

In addition to regulating fuel control, the ECM continually adjusts ignition timing advance. Depending on engine rpm, TPS, MAF, and MAP values, the engine's ignition timing may be advanced or retarded. For example, if a light

These sensors are typical for EFI systems. From top left, clockwise: manifold absolute pressure (MAP) sensor, GM; MAP sensor, Ford; throttle position sensor (TPS), Ford; air temperature sensor; engine coolant (ECT) sensor; O₂ sensor, Toyota; and exhaust valve position (EVP) sensor, Ford.

FUEL INJECTION SYSTEMS

throttle is used in conjunction with moderate rpm, ignition timing is advanced to promote fuel economy. However, if the throttle is wide open but engine rpm is low, timing is retarded to prevent engine ping or knock. The ECM system also uses a knock sensor that "listens" electronically to the engine to determine if the fuel mixture in any of the engine cylinders is detonating, pinging, or knocking. The ECM is able to determine which cylinder is knocking, and then retard ignition timing for only that cylinder, thereby preventing damage to pistons during heavy engine loads.

As you continue your drive, now heading toward the countryside, the vehicle speed sensor (VSS) in the car indicates a steady road speed of 65 miles per hour. The ECM now electronically locks up the automatic transmission's torque converter, creating a solid mechanical connection between engine and transmission. This is why modern vehicles with automatic transmissions get virtually the same gas mileage as standard transmission–equipped vehicles; there is simply no slippage between the engine and transmission, as was the case in noncomputer-controlled vehicles equipped with automatic transmissions.

As you pass through a quaint little town, a group of kids crosses the street. When you come to a stop, the ECM recognizes the throttle has closed based upon inputs from the TPS sensor. The ECM also senses decreasing engine rpm. All of these operating conditions cause the ECM to shut off the fuel injectors, causing a fuel cut (as it is sometimes called). These preprogrammed processes reduce exhaust emissions during deceleration.

While you wait at the crosswalk, you realize how hot it has gotten, so you flip on the air conditioning (A/C). The ECM automatically increases the idle air controller's opening so you don't notice a change in idle speed as the air conditioning compressor turns on. The radiator cooling fan is also controlled by the ECM and is switched on any time the air conditioning is operating. With increased electrical load on the alternator, the ECM increases the IAC signal, which in turn increases engine speed, helping the alternator to keep up with the increased electrical demand from the air conditioning and cooling fan.

Once you reach your destination and shut off your car, the computer goes to sleep and remains this way until you start the process all over again. (The operation of the ECM and its various sensors depicted during our hypothetical drive has been explained in slow motion; in reality, all the electronic signaling and decision-making happens at lightning speed!)

Sensor inputs to the ECM are processed at an amazing rate of over one million times per second. However, the rate at which ECM outputs change is much slower, only about 80 times per second. Because a driver can't open the throttle, turn on the A/C, change gears, or operate any controls faster than an EFI system can, the operation of the engine management system is seamless and undetected by the driver. The end outcome of modern engine management is that it produces vehicles that: 1) start and run well, whether cold or hot; 2) accelerate smoothly; 3) get good fuel mileage; and 4) don't pollute as much as before. As a result, it's not uncommon for some V-8 engines to get over 25 miles per gallon.

SENSOR TESTING
Throttle Position Sensor (TPS)

Most TPS have three wires: power (usually 5 volts), ground, and signal. A TPS can be tested in two ways: by

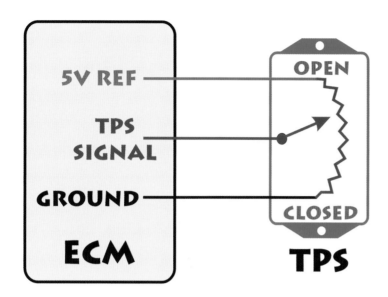

Fig 7-3. *This is a typical TPS/ECM circuit. A TPS is simply a variable resistor. As the throttle is opened, the TPS signal voltage gets closer to the reference voltage of 5 volts. This increasing voltage indicates to the ECM that it should add fuel accordingly as the vehicle accelerates.*

measuring either varying voltage or varying resistance. Measuring TPS voltage is the more accurate method for testing this sensor because the ECM reads its voltage instead of its variable resistance.

To check for the presence of a TPS signal, turn the ignition key to the ON position and leave the TPS plugged in. Back probe each wire with a digital voltmeter. The following readings should be found on each of the three TPS wires: (See figure 7-3)

• **Five-volt reference wire**—The ECM sends 5 volts to the TPS as a reference voltage. If none of the wires at the TPS have 5 volts, check the fuse(s) that power the ECM. If the fuses are good and the TPS wires to the ECM are also okay, there may be a bad power supply within the ECM. If this is the case, the ECM will have to be replaced.

• **Ground wire**—The TPS ground comes directly from the ECM. This wire should read close to 0 volts. If it's higher than 0.02 volt, the ECM may have a bad ground wire. If this is the case, perform a voltage drop test on the ECM ground wire(s) to be sure that is the problem.

• **Signal wire**—The TPS sends a varying voltage signal to the ECM via the signal wire. With the throttle closed, voltage should be around 0.5 volt. As the throttle opens, voltage should gradually increase until it reaches around 4.5 volts at wide open throttle. Inside the TPS is a variable resistor that changes resistance in relationship to throttle angle (opening). This resistor can eventually wear out over many miles of engine operation. Open the throttle slowly while checking the voltmeter reading; a steady increase in

voltage without any skipping or jumping around indicates a good TPS. A DVOM with a bar graph makes it much easier to recognize a bad TPS since a bar graph has a faster display rate than a digital numeric display.

Temperature Sensors

Depending on manufacturer, engine temperature sensors are also referred to as engine coolant (ECT) sensors, coolant temperature sensors (CTS), or other similar acronyms. Intake manifold temperature sensors operate in a similar manner and are often called air charge temperature (ACT), intake air temperature (IAT), or manifold air temperature (MAT) sensors. Both coolant and intake manifold temperature sensors can be tested in the same manner. The majority of temperature sensors are negative temperature coefficient (NTC) thermistors. An NTC thermistor changes its resistance as temperature changes—resistance decreases as temperature increases. There are also a few positive temperature coefficient (PTC) thermistors, but they are rare. These work in a different manner than NCT thermistors—as temperature increases, resistance increases as well.

Temperature sensors are typically equipped with either two or three wires. To check a two-wire coolant or air temp sensor, turn the ignition key to ON and use a DVOM to back probe both wires. One wire should read close to 0 volts since it is the temp sensor ground; the other should have between 0.1 and 4.5 volts depending on its manufacturer and sensor temperature. A cold engine should have approximately 3 volts or higher. Start the engine and watch the voltmeter reading as the engine warms up. Voltage should gradually start dropping to around 1 or 2 volts depending on its manufacturer. If signal voltage drops faster or skips around, the coolant/air temp sensor is probably bad.

TPS sensors come in different shapes and configurations. While most are variable resistors, some Japanese manufacturers use a set, or sets of contact points in addition to the resistor used to provide throttle positions information to the ECM.

Coolant temperature sensors (CTS) containing three wires are typically found on GM vehicles from 1993 or later. The third wire is for the temperature gauge on the instrument panel; the other two wires are the same as described for temperature sensors—ground and signal wires to the ECM. Chrysler vehicles in production from 1985 to 1995 used a dual temperature curve programmed into the ECM. On these vehicles, the ECM uses a set of voltage readings from -20 to 80 degrees Fahrenheit; then it switches internally to a different set of values between 130 and 230 degrees Fahrenheit. If the voltmeter readings jump when the engine's temperature rises to just above 80 degrees Fahrenheit, then the ECM has switched to the "hot" curve—normal for these vehicles. The chart on page 128 lists temperatures, voltages, and resistance values for temperature sensors of domestic manufacturers.

Both types of sensors can also be tested for internal resistance with an ohmmeter after simply unplugging the coolant/air temp sensor. Measuring internal resistance is a good indication of proper function. In general, resistance should be high (thousands of ohms) when cold and low when hot (below 2,000 ohms). See a service manual for exact resistance values for specific vehicles.

Oxygen (O_2) Sensors

The key to making all closed loop feedback systems work is an O_2 sensor, which measures oxygen content in exhaust gas. There are two types of O_2 sensors commonly used—zirconia and titania.

A zirconia dioxide sensor acts as a galvanic battery, generating a small DC voltage based on the comparison between the oxygen content inside the exhaust and surrounding atmosphere. When the oxygen content is low (rich mixture), the difference between exhaust gas oxygen and atmospheric oxygen is high, causing the sensor to produce a high voltage from 0.5 to 0.9 volt. Conversely, when an O_2 sensor detects a lean mixture (high exhaust gas/oxygen content) and compares it to the outside air, the difference is smaller and lower voltages are generated ranging from 0.1 to 0.4 volt.

A titania O_2 sensor operates somewhat differently from a zirconia sensor, but the end result is the same. Instead of producing voltage, a titania O_2 sensor uses a reference voltage from the ECM to change its internal resistance based on oxygen content found in exhaust gas. The resulting voltages will be the same as those for a zirconia sensor. Both sensors have to be hot (600 degrees Fahrenheit) before they can function. Some sensors use an electric heating element to keep them from cooling down at idle (when engine exhaust gas temperature is low). Electronic heating also facilitates quicker sensor warm-up during cold starting.

Unlike other sensors, O_2 sensors can have one, two, three, or even four wires. To test most O_2 sensors, the engine should be warmed to normal operating temperature. To do this, don't

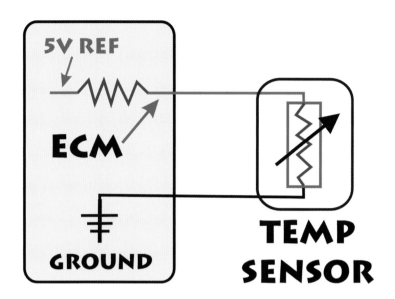

Fig 7-4. The ECM reads the voltage at a point on the temperature sensor's signal wire, located just before the internal resistor (blue arrow). This variable voltage signal is interpreted by the ECM as engine temperature—the hotter the engine, the lower the sensor voltage output.

The sensor on the left is an air temperature sensor, most easily identified by the small windows in its sensing probe. The other sensors are all coolant temperature sensors from various manufacturers.

just let the engine idle in your garage for a few minutes—go for a short drive. Gaining access to the O_2 sensor wires is usually easier at the ECM wiring harness connector than at the sensor itself. Using the positive lead of a digital voltmeter, back probe the O_2 sensor's signal wire at the ECM. Connect the meter's negative lead to a good ground. Start the engine and maintain rpm at 2,000 for 60 seconds. If O_2 voltage starts switching back and forth somewhere between 0.2 and 0.8 volt, the EFI system is in closed loop mode. (For this test, it doesn't really matter if the system is in closed loop or open loop mode.) Next, while watching the voltmeter, snap the throttle open. O_2 voltage should immediately increase to 0.9 volt, indicating a rich mixture. Hold engine rpm steady again, then quickly close

the throttle. This time, O_2 voltage should drop to 0.1 volt or less because the ECM has cut-off fuel to the engine, leaning out the mixture. How fast the O_2 sensor responds to changes in exhaust gas oxygen, as well as the range of voltage displayed (0.1 to 0.9 volt), indicates whether the sensor is good or bad. A good O_2 sensor should make voltage transitions instantly, while a lazy or worn-out one transitions slowly and won't be able to reach 0.9 volt no matter how rich the fuel mixture becomes.

MAP and BARO Sensors

Manifold absolute pressure (MAP) sensors are nothing more than electronic vacuum gauges connected to a

FORD CTS

Degrees F	Degrees C	Volts	Ohms
248	120	0.28	1,180
230	110	0.36	1,550
212	100	0.47	2,070
194	90	0.61	2,800
176	80	0.80	3,840
158	70	1.04	5,370
140	60	1.35	7,600
122	50	1.72	10,970
104	40	2.16	16,150
86	30	2.62	24,270
68	20	3.06	37,300
50	10	3.52	58,750

CHRYSLER CTS HOT CURVE

Degrees F	Degrees C	Volts	Ohms
230	110	1.80	
210	99	2.20	640–720
200	93	2.40	
190	88	2.60	
170	77	3.02	
150	66	3.40	
130	54	3.77	

CHRYSLER CTS COLD CURVE

Degrees F	Degrees C	Volts	Ohms
80	27	2.44	9K–11K
70	21	2.75	
60	16	3.00	
50	10	3.30	
40	4	3.60	
30	-1	3.90	29K-36K

GM CTS

Degrees F	Degrees C	Volts	Ohms
210	100	0.8	185
160	70	1.5	450
100	38	3.0	1,800
70	21	4.4	7,500
20	-7	4.6	13,500

These charts list voltage and resistance values for coolant temperature sensors. Air temperature sensor charts are similar.

vacuum source at the engine's intake manifold. MAP sensors are either bolted directly to a manifold or at a remote mounting location if a vacuum hose is used. A MAP sensor provides the ECM with engine load information, basically letting the ECM know when the engine is working hard. GM and Chrysler MAP sensors send a variable voltage to the ECM; Ford MAP sensors send a varying frequency to the ECM.

Whenever the engine is idling, the negative pressure inside the intake manifold is high—around 20 inches of mercury. Once the throttle is opened, the vacuum (negative pressure) drops, until, at wide-open throttle, engine vacuum is at 0-inch Hg inches of mercury. The MAP sensor senses the engine vacuum, and then outputs a voltage/frequency signal, which it sends to the ECM for processing.

Similar in operation to a MAP sensor, a BARO sensor measures ambient air pressure (altitude) and sends a signal to the ECM. The ECM changes fuel and timing parameters depending upon the altitude the vehicle is being operated at. Mountain driving requires a leaner fuel mixture, so the ECM changes the injector pulse width accordingly.

All MAP sensors have three wires: power (usually 5 volts), ground, and signal. Following is some information on the different domestic manufacturers of MAP sensors.

GENERAL MOTORS

Many GM cars and trucks are equipped with either a MAP, BARO, VAC (vacuum), or turbo MAP sensor. They all may look similar, but each can be identified to some extent by the color of the female plug on the sensor. Black or gray colored plugs are found on VAC sensors. MAP or BARO sensors have colored plugs (but not black, gray, or orange), with the exception of a turbo MAP, which uses an orange plug. All sensors have black plastic outer cases.

To test GM sensors, turn the ignition key to ON. Using a DVOM, back probe all three wires. One wire should have 5 volts (power), one should be close to 0 volts (ground), and the last should be the signal wire.

Use a hand vacuum pump to test MAP and VAC sensors. As vacuum pressure to the sensor is varied, the voltage should change. MAP sensor voltage must be adjusted for testing above sea level since voltage readings decrease about 0.2 volt every 1,000 feet above sea level. To find voltage values for each type of sensor, see the chart on page 131.

FORD MOTOR COMPANY

Ford MAP sensors produce a frequency output instead of voltage output. Like a GM MAP sensor, a Ford MAP sensor

From left to right are O$_2$ sensors with one, two, three, or four wires. O$_2$ sensors with three or four wires also have internal heating elements that keep the sensors at operating temperature, even if the exhaust gas temperature is low due to prolonged engine idling.

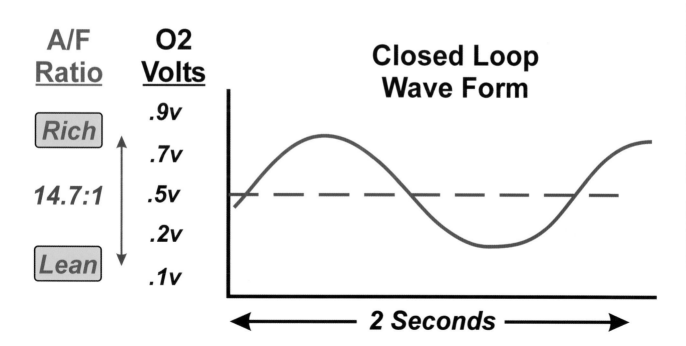

Fig 7-5. *This waveform shows an O$_2$ sensor switching between rich and lean air/fuel mixtures. Rich mixtures are indicated by sensor voltage readings above 0.5 volt, while readings below 0.5 volts are for lean air/fuel ratios.*

All MAP sensors are basically electronic vacuum gauges. This Ford MAP sensor (on the left) sends a variable frequency to the ECM. Other MAP sensors change their voltage signals to the computer to indicate engine load.

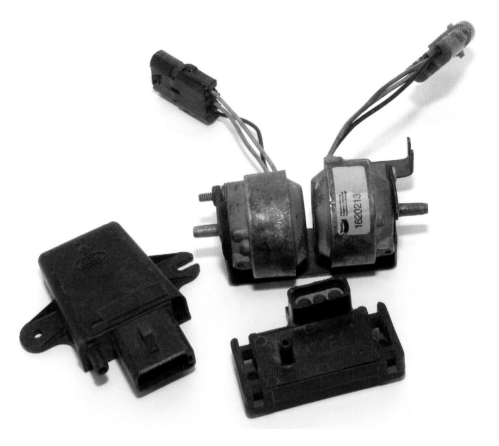

has three wires. To test this sensor, use a DVOM to back probe each of the three wires while the MAP sensor is plugged in. With the ignition key turned to ON, the power wire should read 5 volts, the ground wire should be close to 0 volts, and the signal wire to the ECM should read 2.5 volts. (Only a multimeter capable of reading frequency can be used to check the signal wire.) Next, use a hand vacuum pump to change vacuum levels at the MAP sensor. Voltage on the signal wire should always be 2.5 volts, regardless of the vacuum level at the MAP. By contrast, frequency (hertz) levels should change as vacuum levels change at the sensor. However, just like on other MAP sensors, when checking this sensor above sea level, frequency readings will be slightly lower—about 3 hertz for every 1,000 feet above sea level. To find frequency values for vacuum levels, see the chart on page 132.

If no meter is available to read frequency, another quick way to check if the MAP sensor is working requires use of a jumper wire. Attach the alligator clip of a jumper wire to the signal wire (usually the middle wire) at the MAP sensor. Connect the other end to the antenna of the vehicle being tested. If the antenna is embedded in the windshield, try holding the clip against it. Then turn the ignition to ON and tune the radio to an AM frequency below 600. (Don't expect good tunes here!) The map sensor's frequency output will come right through the radio speakers. Again, use a

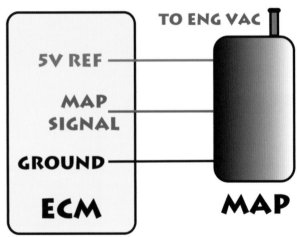

Fig 7-6. *MAP sensors have three wires: power (5-volt reference), ground, and signal. Some MAP sensors have a nipple fitting to accept a vacuum hose, which connects the sensor to the intake manifold.*

hand vacuum pump to change vacuum levels at the MAP sensor. The sound of the pitch of the frequency should change, confirming the MAP sensor is working (but unfortunately, not giving enough information to measure its accuracy). In the Ford MAP chart, Hg represents inches of mercury and kPa represents kilopascal, a unit of pressure.

This GM MAP sensor has a nipple connecting the sensor (via a vacuum hose) to the engine intake manifold. These sensors produce a voltage signal to the ECM. Courtesy CARQUEST Auto Parts

GM MAP	
Vacuum	Voltage
0	4.86
5	3.96
10	3.06
15	2.10
20	1.10
21	0.82

GM BARO

3.00 to 4.86 depending on altitude

GM TURBO MAP	
Vacuum	Voltage
0	2.40
5	1.66
10	1.18
15	-0.70
20	0.26

GM VAC	
Vacuum	Voltage
0	0.54
5	1.38
10	2.20
15	3.20
20	4.20
21	4.40

These are vacuum versus voltage values for GM's MAP, vacuum (VAC), and barometric pressure (BARO) sensors. These numbers serve as a general guide when diagnosing MAP sensors; see a service manual for specific voltage/vacuum values.

CHRYSLER

A Chrysler MAP sensor is similar to a GM sensor in that both produce a varying voltage that changes with engine vacuum levels. Chrysler uses two types of MAP sensors—turbo and nonturbo. Both work the same, but voltage outputs are different. On some early pre-1987 vehicles, the MAP sensors are located inside the vehicle's logic module of the onboard computer and are therefore not accessible for testing. If the MAP sensor has to be replaced, there is, fortunately, a relocation kit available for this system.

To test a Chrysler MAP sensor, turn the ignition key to ON. Using a DVOM, back probe all three wires—one should have 5 volts (power), one should be close to 0 volts (ground), and the last should be the signal wire. Use a hand vacuum pump to test the MAP sensor. When testing above sea level, voltage readings run about 0.1 volt lower for every 1,000 feet above sea level. To find voltage values for each type of Chrysler MAP sensor, see the chart on page 132.

MASS AIRFLOW (MAF) SENSORS

A MAF sensor measures the volume of air entering an engine, calculates air density and temperature, and sends a signal of the result to the ECM. A MAF sensor on a GM or Ford car is often referred to as a hot wire MAF sensor because the temperature of its sensing element is maintained to a programmed level above that of incoming air. As air passes over the sensing element, it cools down, requiring the MAF sensor to produce more current to maintain the programmed temperature. By measuring the internal current flow used to heat the sensing element, the small computer inside the MAF can calculate mass airflow into the engine. The digital signal output produced is a varying frequency, which the ECM uses to calculate injector pulse width and ignition timing.

GM MAF Sensors

GM uses three types of MAF sensors: AC Delco, Bosch, and Hitachi. The AC Delco three-wire MAF sensor was installed in cars manufactured between 1985 and 1991. To test this type of sensor, leave it plugged in with the ignition key in the ON position, and then back probe each wire using a DVOM. The power wire should read 12 volts, the ground wire should be close to 0 volts, and the signal wire should have 2.5 volts. (Remember, the multimeter must be capable of reading frequency or hertz to measure this sensor's output.) With the key turned to on and the engine off, the signal wire readings should be around 8 to 10 hertz, while at warm idle they should be around 40 to 45 hertz, and at wide-open throttle about 150 hertz. These numbers are approximations, but as long as the hertz readings increase when the throttle is opened, the MAF sensor is probably good. Also, since these sensors often have bad wiring harness connectors, be sure to check the connector before condemning the MAF as bad.

AC Delco five-wire MAF sensors were installed in cars manufactured between 1985 and 1989. This sensor's connector has letters identifying each terminal. Following is a list of functions for each of the wires used on this sensor:
• Terminals A and B have a black/white wires and are ground.
• Terminal C has a dark green wire that is the signal wire and should have 0.5 volt with the engine idling. To determine if the sensor is good, change the function on a DVOM to read frequency. This wire should have 32 hertz at idle and 150 hertz at wide-open throttle.
• Teminal D is a dark blue wire and receives 12 volts during burn-off cycle. The burn-off cycle keeps the hot wire clean by burning off any dirt or contaminants; this occurs whenever the ECM sends a signal to the MAF to heat the sensing wire to red-hot. (This occurs automatically and is controlled by the ECM.)
• Terminal E is a red or purple wire and also receives 12 volts to power the MAF. However, be careful when checking this wire for voltage with the key turned to ON and the engine not started, as there are 12 volts on this wire for only two seconds and it's easy to miss the meter reading. These sensors are also prone to dirty hot-wires, despite their burn-off cycle. The wire can be cleaned using a Q-tip and rubbing alcohol and dried by gently blowing across it with your mouth. Since the wire is very thin, it breaks easily, so don't use an air nozzle to blow it off.

The Hitachi MAF sensor was used from 1988 to 1993 on GM 3300 and 3800 engines. This sensor also has three wires. To test this sensor, leave it plugged in, turn the ignition key to ON and then back probe each wire using a DVOM. The power wire should have 12 volts, the ground wire should be close to 0 volts, and the signal wire should have 5 volts. Again, the multimeter must be able to read frequency to measure this sensor's output. With the ignition key ON, and the engine off, the reading should be between 0 and 35 hertz, and at idle it should be between 2,500 and 3,500 hertz. As long as the hertz readings go up when the

FORD MAP		
VAC Hg	kPa	Hertz
0	0	159
3	10.2	150
6	20.3	141
9	30.5	133
12	40.6	125
15	50.8	117
18	61.0	109
21	71.1	102
24	81.3	95

Ford MAP sensors produce frequency instead of voltage outputs. A DVOM capable of reading frequency, or hertz, is necessary to test this type of sensor.

CHRYSLER MAP	
Vacuum	Voltage
0	4.5
5	3.7
10	2.9
15	2.1
20	1.2
25	0.4

CHRYSLER TURBO MAP	
Vacuum	Voltage
0	2.25
5	1.85
10	1.45
15	1.05
20	0.60
25	0.20

Chrysler's turbo MAP has different voltage values from a nonturbo MAP. When ordering a new sensor, do not get them mixed up; your vehicle will not run well, or at all, with the wrong sensor.

FUEL INJECTION SYSTEMS

A MAF sensor's hot wire is very thin, but it is protected by a plastic/metal housing and screen on both ends of the sensor. When incoming air cools the hot wire, the MAF computer increases current to keep it at a programmed temperature.

throttle is opened, the MAF sensor is most likely good. This sensor's hot wire can also be cleaned in the same manner as a five-wire AC Delco MAF sensor.

Ford MAF Sensors

Ford mass airflow sensors are similar to those used on GM vehicles, except these sensors produce voltage instead of frequency. The following test is provided for 1990 to 1995 Ford MAF sensors equipped with four wires.

To test this sensor, leave it plugged in, turn the ignition key ON, then back probe each wire using a DVOM. The power wire should have 12 volts. There are two ground wires that should each be close to 0 volts. The signal wire should also be close to 0 volts. At idle, the signal voltage should be about 0.8 to 1.0 volt, but at 3,000 rpm the reading should increase to about 2 volts, and when the throttle is snapped wide open, the reading should be about 3.5 volts. As long as the voltage increases when the throttle is opened, the MAF sensor is probably good. This sensor is also prone to having a dirty sensing wire, but it can also be cleaned with rubbing alcohol and a Q-tip. Be sure to dry it very gently by blowing with your mouth, since this wire is also very thin and easily broken. (Don't use an air nozzle!)

Knock Sensor (KS)

A knock sensor (KS), or detonation sensor as it is sometimes called, has a piezoelectric crystal that generates voltage when subjected to mechanical stress. This crystal produces an electrical signal with a unique signature based upon engine knock or ping. The KS operates similarly to an electronic lighter for a gas refrigerator or barbecue, except it doesn't spark; rather, it only produces voltage. Whenever an engine knocks or pings under heavy acceleration, the KS sends a signal to the vehicle's ECM. The ECM then retards ignition timing in an effort to stop the knocking—a safety feature that can prevent damage to pistons and rings. The use of a KS also allows for the use of different grades of gasoline without engine damage or poor performance.

The following are general diagnostic tests for KS, but service manuals should be consulted for specific tests, as these sensors sometimes vary in their outputs.

To test this sensor, disconnect it and probe the sensor wire with a DVOM. Set the multimeter to read AC millivolts. Take a small hammer and tap on the engine block (near the sensor) while watching the voltmeter. The sensor should produce a small amount of AC voltage, usually less than 4 volts.

Another method for testing KS operation involves use of an ignition timing light. This method only works on older computer-controlled vehicles—usually those with ignition distributors. With the engine running and the timing light connected to the number one cylinder's spark plug, ignition timing can be checked at the crankshaft pulley. Use a small hammer to tap the engine block near the KS. The ignition timing should retard a few degrees, proving the KS is sending a signal to the ECM, which is then retarding the ignition timing. This test won't work on all vehicles, so be sure to check a service manual for specific tests.

Vehicle Speed Sensors (VSS)

The VSS provides a signal to the ECM to indicate vehicle speed. The ECM uses this information to control the automatic transmission torque converter lock-up and to determine shift points. There are three types of these sensors: (1) an AC pickup coil, (2) a hall-effect switch, and (3) an optical sensor. All three types of speed sensors operate in a similar manner as their ignition counterparts do in ignition systems. (See Chapter 6 for an overview of testing methods used on speed sensors.)

Idle Air Control (IAC)

IAC sensors are not really sensors at all—they are, in fact, actuators. The ECM controls these devices in order to change engine idle speed. Idle speed is increased by the ECM during cold start conditions and air conditioning operation, or whenever the automatic transmission is shifted between PARK and DRIVE. Certain IAC systems use a stepper motor to control the amount of air allowed into the engine, while others use a bypass valve. In general, these actuators can be checked for proper resistance, but are difficult to test for actual operation. Using a computer scan tool (hand-held computer interface) is sometimes the only way to verify if the ECM and IAC actuator are functioning together correctly. Consult a service manual for proper testing procedures for the specific vehicle being worked on.

FUEL INJECTORS AND INJECTOR PULSE

Fuel injectors are simply nothing more than actuators controlled by a vehicle's computer. The ECM controls the amount of fuel injected into the engine by varying the pulse width of the injectors. How long the injectors are open (allowing fuel to be injected) is a function of the corresponding length of time the injector pulse width lasts.

All fuel injectors are solenoids consisting of a coil of wire and moveable electromagnetic valve. When energized, the coil of wire acts like a magnet, causing the valve to open.

This AC Delco MAF sensor has been cut in half so the computer is visible (left). The hot wire is located inside the round plastic housing. All of the air entering the engine must first pass through the sensor so it can be measured.

A fuel injector's resistance can be measured using an ohmmeter. However, resistance readings only confirm that an injector's internal coil is not shorted, or open. Because resistance values vary widely between injector manufacturers, consulting a service manual for specific values for the particular vehicle being worked on is always a good idea. The only way to tell if an injector is actually working electrically is to use a lab scope to monitor the injector's pulse from the ECM. However, a number of simple tests will work most of the time to confirm indirectly that an injector is operating. The following first four tests verify the ECM is sending an injector pulse to the injector, while the last test checks for mechanical operation.

1. Unplug the injector and connect a test light between the two wires at the injector harness connector. Crank or start up the engine while watching the test light. If the light flashes, the ECM is sending an injector pulse to the injector. (Although a test light works most of the time for this test, it's important to know this test will not work on vehicles that use a dropping resistor in the injector circuit, thus limiting current going into the injector to prevent overheating.)

2. Instead of using a test light, use a noid light specific to the particular EFI system tested. A noid light has low enough resistance to flash during a test, even when a dropping resistor is used. (See Chapter 3 to find information about noid lights.)

Every IAC receives a signal from the vehicle's ECM to adjust engine idle speed. Some controllers are stepper motors, while others are bypass valves. The function of an IAC is to bypass the air around the throttle plates in order to increase engine idle speed. Courtesy CARQUEST Auto Parts

This VSS (vehicle speed sensor) tells the ECM to lockup the torque converter clutch. Courtesy CARQUEST Auto Parts

3. Use an inductive ignition timing light to verify injector pulse. Clamp the timing light's probe around any wire going to the injector. Start or crank the engine and watch the timing light to see if it flashes. A flashing light provides confirmation that the ECM is sending an injector pulse to the injector.

4. Use a logic probe to verify the presence of an injector pulse signal from the ECM. Connect a logic probe or red-and-green test light to the vehicle's battery. (See Chapter 3 for more information about logic probes and test lights.) Touch the probe to an injector wire (the injector can be left plugged into its connector). Crank or start the engine and watch the LED on the logic probe. If it flashes or pulses, then the ECM is sending an injector pulse to that injector.

5. For a low-tech method, simply take a long screwdriver and touch either end to the fuel injector and stick the other end in your ear (no kidding!). If the injector is working, you'll hear a steady clicking from the injector. A wooden dowel or mechanic's stethoscope works just as well for this test.

While they may look different, all of these fuel injectors operate in the same manner. Each uses a coil of wire that acts like a magnet when the ECM sends out an injector pulse. The length of the pulse (pulse width) determines how long the injector remains open and, consequently, how much fuel is injected into the engine.

YEL-LT.BLUE
T.GRN-YEL
BLACK
LT.GRN-BLK
BLACK
WHITE-RED
LIGHTS
(4 USED)
HIGH BEAM
INDICATOR
DOME
BLACK

SECTION IV
GENERAL ELECTRICAL SYSTEM DIAGNOSIS

CHAPTER 8
WIRING DIAGRAMS

Taking a cross-country car trip when you were young might have been your introduction to a road map. Maybe your parents showed you a road atlas and said something like, "We're here now, and after driving for four days we'll be there." Your first introduction to reading a wiring diagram was undoubtedly less memorable. However, reading a wiring diagram is similar to reading a road map. Road maps illustrate how to get from point A to point B. In a sense, so do wiring diagrams. However, instead of connecting interstates, highways, and roads, a wiring diagram shows interconnected major electrical systems, subsystems, and individual circuits. When a technician looks at a wiring diagram, the goal is to figure out why a circuit isn't operating. Consequently, destination points are replaced with power sources, controls, identification of and routes to load devices, and pathways for ground returns back to the negative battery terminal.

Wiring diagrams and road maps also have another feature in common—layers of detail. For example, if you look at a road atlas of California, you won't be able to locate a street address. You may locate a particular city or town, but you won't find a specific route to an exact address. To find the exact location of a particular residence or destination, you need a detailed street map. The same is true (though to a lesser extent) of wiring diagrams.

Originally, wiring diagrams showed the electrical systems for an entire vehicle. After the early 1990s, wiring diagrams were categorized and separated into the major electrical systems and their subsystems (not the entire vehicle). This occurred, in part, due to the complexity of late model vehicles; with so many electrical components, it was no longer practical to put everything on one wiring diagram. Today, there are numerous wiring diagrams for each vehicle, each pertaining to specific electrical systems or subsystems in which particular electrical components belong. For example, if a dome light inside a vehicle isn't

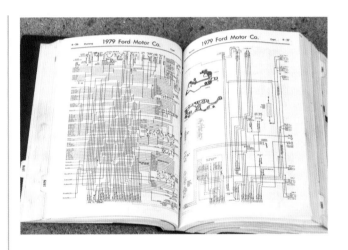

Many automotive technicians with years of experience prefer to use a book version of wiring diagrams. However, today, most wiring diagrams intended for professional use can be found on a CD or DVD. A technician can read them on a computer monitor and print out, one page at a time, as needed. Courtesy Scott's Speed Shop and Mitchell 1.

receiving 12 volts when a passenger door opens and you want to find out why, it can take several pages of wiring diagrams to map out the exact path of power from the battery to the light. The diagram for interior lighting will most likely show you how the interior wiring and fuses are connected, but the power to the interior fuse panel may originate with an under-hood relay or fuse box. Consequently, a different wiring diagram may be needed to determine where the relay receives power.

The secret to reading a wiring diagram will not be found on the page that identifies the electrical symbols used. While this information is certainly valuable, it won't really tell you how to read a wiring diagram. The secret to deciphering a wiring diagram is, quite simply, understanding how a circuit and/or load device operates. If you

L-LT.BLUE
GRN-YEL
BLACK
BLACK
LT.GRN~BLK
BLACK
WHITE-RED
LIGHTS
(4 USED)
HIGH BEAM
INDICATOR
DOME
BLACK

25a
Fuse
Dash
Sw
Drive Lts
Battery

25a
Fuse
Dash
Sw
Drive Lts
Battery

Fig 8-1. *This multicolor wiring diagram shows a driving light circuit in both on and off conditions. Unfortunately, most manufacturers don't provide the luxury of color diagrams; they are typically in black and white only.*

keep in mind the "Three Things" (introduced in Chapter 1) that all 12-volt DC circuits must have in order to operate (power, ground, and a load device), cracking the mystery of a complex wiring diagram will be much easier.

Every load device (motors, bulbs, relays, solenoids, computers) requires both a power source and ground return and must be controlled. Some load devices are switched on and off via a power source, while others are controlled by switching ground returns on or off. Load devices are sometimes dependent on other load devices to operate, while some produce signals used by solid-state electronics to trigger other load devices as well. Determining how a specific load device is controlled can usually best be accomplished by consulting a wiring diagram.

This can get complicated, so to make it easier, we'll start with some basic examples of circuit wiring diagrams and then add layers of complexity as we go. Understanding how these sample diagrams relate to the circuits they depict should help with the task of reading more complex vehicle wiring diagrams later on.

READING WIRING DIAGRAMS

Figure 8-1 is a simple diagram showing a driving light circuit in both on and off positions. Its circuit consists of a 25-amp fuse (used to protect the circuit), a switch (located on the dash of the vehicle), and two driving lights. The red-colored wires have 12 volts, the black wires are ground returns (0 volts), the purple wires represent load devices, and the yellow areas are intended to show the lights operating. With the dash switch in the off position (upper circuit), the power wire from the battery to the switch has 12 volts. When the dash switch is in the on position (lower circuit), the wires to both driving lights have power and the

lights are on. The ground wires at each light have 0 volts, since all available voltage is used by the lights.

Reading the wiring diagram illustrated in Figure 8-1 is easy for two reasons. The driving light circuit has been isolated and isn't shown as a subcircuit of the overall lighting system; and the wires and load devices are different colors—red for power, black for ground, and purple for load devices (bulb filaments). Unfortunately, most wiring diagrams do not provide any of these advantages. Even late model vehicle diagrams may not isolate driving lights to this extent—more likely they will be depicted as part of the overall lighting system. Furthermore, color enhancement, if used at all in a wiring diagram, is solely for the purpose of identifying individual wires, not for indicating power or ground sides of a circuit. Most importantly, wiring diagrams will not show the differences between a circuit in its on and off states.

Consequently, the circuit depicted in the sample wiring diagram illustrated in Figure 8-1 does not truly represent automotive reality. It is too simplistic. However, the basic inherent problem with its design can be solved simply by the addition of a single component—a relay. Here's why: In a real automotive application, a relay would be used instead of a heavy-duty switch and wires to provide the high-amperage connection needed between the driving lights and battery. A dash switch would still be part of the overall circuit, but now it only has to switch the relay on (a much smaller amperage load), instead of the driving lights. Consequently, the switch and the wires to and from it don't have to be heavy-duty since the relay, instead of the switch, is turning the lights.

Figure 8-2 shows a relay added to the driving light circuit. The relay connects battery power directly to the driving lights; it is switched on and off via a dash switch.

WIRING DIAGRAMS

137

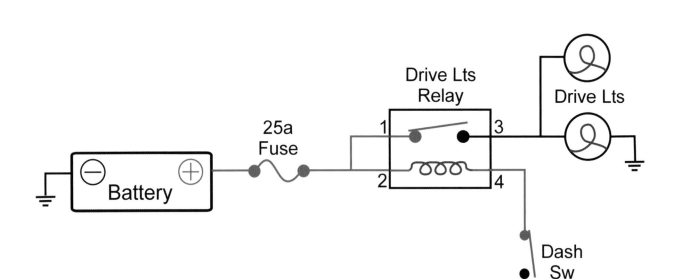

YEL-LT.BLUE
LT.GRN-YEL
BLACK
LT.GRN-BLK
BLACK
WHITE-RED
BLACK
LIGHTS
(4 USED)
HIGH BEAM
INDICATOR
DOME
BLACK

Fig 8-2. *A relay has been added to the driving light circuit. The relay now controls the high amperage load that the driving lights need to operate (instead of the switch and wires depicted in Figure 8-1). The addition of a relay allows for the use of smaller-gauge wires on the dash switch since it now controls only the relay, instead of directly controlling the driving lights. In fact, the switch itself can be made smaller still, as it no longer has to contend with a high electrical load with this design.*

Because the control coil inside the relay is low amperage (about 1 amp), the dash switch and its wires don't have to be too heavy-duty. The relay contains two circuits: 1) a low amperage circuit, consisting of a single control coil, that causes the relay's contact switch to close when energized; and 2) a high amperage circuit that acts as a switch to connect the driving lights directly to the battery's positive terminal. The use of relays in late model vehicles is very common, so understanding how to figure out if one is working or not will help solve many electrical problems.

To better understand the driving light circuit in the wiring diagram of Figure 8-2, each subcircuit must first be identified and then isolated from the larger driving lights circuit. In this diagram there are two separate circuits and control switches: 1) a driving lights circuit with high-amperage relay contact switch; and 2) a relay control-coil circuit and dash switch. Each circuit requires power, load device, and ground return. In addition to the essential "Three Things," each circuit also has a switch controlling it. However, in this case, a dash switch now controls a relay instead of controlling the driving lights, a much smaller electrical load. Consequently, the driving lights are now connected to the battery via contact points inside the relay instead of directly through the dash switch. In addition, inside the relay is a control coil (used to close the relay contacts), which is connected to ground via the dash switch inside the vehicle. Therefore, the wires going to and from

Automotive relays come in all shapes and sizes, but they all basically operate the same. Some late model vehicles use as many as 30 relays or more.

the control switch for the driving lights can now be quite small since they will carry only the amount of energy required to power the relay's control coil—less than 1 amp. By contrast, the contact switch inside the relay now carries the high amperage load needed to power the driving lights.

When the dash switch is turned to the ON position, it connects a ground to the relay's control coil (terminal 4 on the relay in the diagram of Figure 8-2). The control coil receives power directly from the battery. Since the control coil is now on (it has power and ground), it develops a magnetic field, which in turn pulls the high-amperage relay contact switch down, causing relay terminals 1 and 3 to connect (see Figure 8-2). With terminals 1 and 3 connected, the driving lights are able to receive battery power and turn on.

To better understand how the relay operates, an examination of the voltage present on the relay's terminals with the driving lights' circuit in both an off and on state is necessary. Unfortunately, the diagram in Figure 8-2 is typical of real wiring diagrams, in that the voltage value for each wire and relay terminal must be imagined in both circuit conditions since both states are not shown in the diagram. The diagram will never show the circuit in its on state; this makes using this diagram a little more challenging than the diagram shown in Figure 8-1. The wiring diagram exhibited in Figure 8-2 is pretty typical of those distributed by vehicle manufacturers (with the exception of the added feature of colored wires indicating power and ground).

Figure 8-2 is a working example of a wiring diagram. Measuring voltage levels at various points in the diagram will illustrate how the circuit works. With the dash switch in the OFF (or open) position, voltage at the relay terminals shown in Figure 8-2 will be as follows when measured with a voltmeter:

• Terminals 1, 2, and 4 will all have 12 volts. Terminal 4 has 12 volts because the dash switch is "open" (not connected).
• Terminal 3 has 0 volts (ground).

However, when the dash switch for the driving lights is in the ON or closed position, the voltage readings at the same relay terminals change as follows:

• Terminals 1 and 2 remain unchanged at 12 volts.
• Terminal 3 now reads 12 volts (since the driving lights are on).
• Terminal 4 is now at 0 volts instead of 12, because it is now acting as a ground return for the relay's control coil circuit.

Continue to reference Figure 8-2 as a hypothetical wiring diagram, and use a voltmeter assuming the problem is the driving lights don't work. Following are a sequence of steps that should be performed in order to discover the cause of this problem.

1. With the dash switch in the OFF position, relay terminals 1 and 2 should each have 12 volts when touched by a voltmeter probe. These readings confirm the 25-amp fuse is good and so are the associated wires going to the relay.

2. Voltage at relay terminal 3 should be 0 volts. (No voltage is expected at terminal 3 because the dash switch is in the OFF position.)

3. Relay terminal 4 is a different matter. This terminal should have a reading of 12 volts when the switch for the

Typical automotive relays have two components: the control coil circuit (low amperage) and a set of high amperage contact points, which act like a switch used for connecting load devices to a power source. The high amperage contacts are located on the lower left. The copper wires located on the relay control coil are visible in the center of the relay.

driving lights is in the OFF position, as it connects directly to the battery through the relay control coil. Instead, terminal 4 has a reading of 0 volts. This indicates the relay control coil is open (has a broken wire) because there is battery voltage at terminal 2.

4. Using a jumper wire, you can connect relay terminals 1 and 3 together. This connection serves to bypass the relay completely, thus removing it from the circuit. As a result, the driving lights should come on. This confirms that the power wire from the relay to the driving lights is okay, and the driving lights ground wire is good as well.

5. The last step is to verify whether the dash switch is connected and working. Connect the red (positive) voltmeter lead to relay terminal 2. Then connect the negative lead to relay terminal 4. Once the switch is moved to the ON position, the meter should display a reading of 12 volts, indicating the switch is both good and connected to ground.

As can be seen, adding components to an imaginary electrical system adds corresponding detail to the sample wiring diagram; thus, the wiring diagram in Figure 8-3 is actually starting to look more like a real-life wiring diagram found in a vehicle service manual.

In Figure 8-2, the addition of a relay solved the design problem of having to use large gauge wires and a heavy-duty

YEL-LT.BLUE
.T.GRN-YEL

BLACK

BLACK
LT.GRN-BLK
BLACK
WHITE-RED

LIGHTS
(4 USED)
HIGH BEAM
INDICATOR

DOME

BLACK

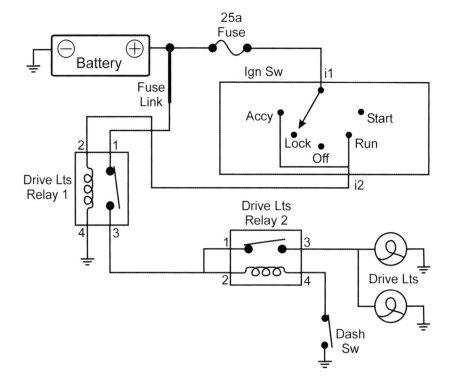

Fig 8-3. *Two relays are used in this driving light circuit. Relay 1 prevents the driving lights from being left on if the ignition key is removed, even if the dash switch is left on.*

switch in the driving light circuit (see Figure 8-1). However, the diagram in Figure 8-2, though closer to automotive reality, still has a design flaw. There is no protection against a driver leaving the driving lights on without realizing it, a mistake that will cause the battery to go dead after only a few hours.

The wiring diagram depicted in Figure 8-3 shows how the addition of another relay controlled by the vehicle's ignition switch prevents the driving lights from being left on by mistake. The purpose of relay 1 is to connect power to relay 2 (the original relay depicted in Figure 8-2). Relay 2 can only receive 12 volts when the ignition key is in the ACC (accessory) or RUN positions. If the key is in the START or OFF position or removed from the ignition completely, no power is available at relay 2, preventing the driving lights from being left on inadvertently, even if the dash switch is left on. An additional benefit of this improved circuit design is that the driving lights also can't operate during engine startup, and with the driving lights off, more amperage is available from the battery to start the engine.

With all of the newly added features to the basic sample circuit design, it now operates more like a typical automotive electrical circuit. Driving light relay 2 operates in the same manner as the original relay in the wiring diagram in Figure 8-2. As long as it has power, the driving lights can be switched on or off via the dash switch. The ignition switch controls relay 1. With the ignition switch in the ACC or RUN position, power becomes available at terminal 2 of relay 1.

However, terminal 4 is a constant ground, and with power at terminal 2, the relay's high amperage contact switch closes. This action connects terminals 1 and 3, sending power to relay 2. As a result, terminal 1 on relay 1 receives power directly from the battery via a fusible link. Finally, terminal 3 of this same relay is connected to terminals 1 and 2 on relay 2; as mentioned earlier, this relay operates in the same manner as the original relay in Figure 8-2.

The sample wiring diagram in Figure 8-3 is finally starting to look more like a real wiring diagram found in a vehicle service manual. To diagnose a problem within the circuit that may result in nonoperational driving lights, you need to follow the flow of power from the battery to the driving lights. To check this, the following terminals should be "hot" with the ignition key in the RUN position and the driving light dash switch in the off position:
• Terminals i1 and i2 (ignition switch)
• Terminals 1, 2, and 3 of driving light relay 1
• Terminals 1, 2, and 4 of driving light relay 2 (All of these should produce a voltmeter reading of 12 volts)
However, when the dash switch is turned to the on position, the following readings should occur:
• Terminal 4 of relay 2 should read 0 volts
• Terminal 3 of relay 2 should be "hot" (or 12 volts)
If the lights still don't operate, there is a possibility of a break in the wire leading from terminal 3 of relay 2 to the

L-LT.BLUE
GRN-YEL

BLACK

BLACK
LT.GRN-BLK
BLACK
WHITE-RED

LIGHTS
(4 USED)
HIGH BEAM
INDICATOR

DOME

BLACK

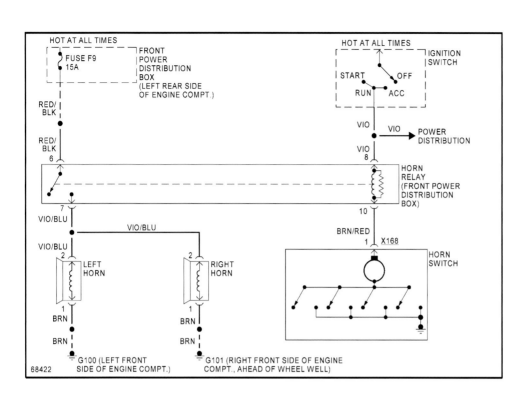

Fig 8-4. *This wiring diagram isolates a typical horn circuit. To locate the source of power to the ignition switch and front power distribution box, another diagram will have to be consulted.*
Courtesy of Mitchell 1

driving lights. Otherwise, the bulbs may be burned out or the ground wire for both lights may be broken.

Figure 8-4 is an actual automotive wiring diagram for a 1993 BMW 525i horn circuit. This is an example of a wiring diagram that has been separated, or broken down, into systems. In addition, all wires are identified by wire color.

It will take several steps to read this wiring diagram and analyze how this horn circuit works:

1. Starting at the top left portion of the diagram, fuse F9 is a 15-amp fuse that powers one-half of the horn relay on a red/black wire. The diagram provides additional information about fuse F9, including its condition (hot at all times) and location (front power distribution box, left rear side of engine compartment). If the horns didn't work and fuse F9 wasn't receiving power, another diagram would have to be consulted to determine the power source for the Front Power Distribution Box.

2. Power for the horn relay control coil comes from the ignition switch via a violet wire. Again, a separate diagram would have to be examined if no power was present on the violet wire from the ignition switch to the relay. With the key in the START, RUN, or ACC positions, horn relay terminals 6, 8, and 10 should all have 12 volts.

3. The horn switch acts as a ground for terminal 10 of the horn relay via a brown/red wire. When this wire makes contact with a ground, the relay's contacts close, thereby connecting terminals 6 and 7 and completing the horn

circuit. The horn switch is located in the center of the steering wheel and has four contacts connecting the brown/red wire to ground.

4. If both horns in the diagram don't work and you want to bypass the horn relay as a means of eliminating it as the source of an electrical problem, you should connect a jumper wire between relay terminal 6 (red/black wire) and terminal 7 (violet/blue wire) to make them operate. Just be sure to hold your hands over your ears to prevent hearing loss!

WIRING DIAGRAM EXERCISES

The most effective way to learn how to read and use a wiring diagram is by practicing. With that in mind, the next several wiring diagrams are followed by a series of questions for you to answer. Write your answers down on a piece of paper as you look at each diagram. The correct answers to the questions and their analysis can be found at the end of this chapter. When taking tests in school, some of your teachers probably told you "there are no trick questions." Such is not the case here, as there are a few trick questions designed to make you think! Good luck!

Figures 8-5 and 8-6 show an actual wiring diagram for a 1975 Ford Thunderbird. The engine in this car is carbureted—no EFI, but it does have an electronic ignition. The entire wiring diagram for this vehicle is on only two pages. Reading from left to right, the following circuits are found on the first page: front lighting, ignition system, battery,

141

charging system, and fuse box. The second page includes windshield wiper switch, turn signals, instrument lights, ignition switch, light switch, door switches, and rear lighting. In 1975, the Ford Motor Company (in its wisdom) decided that wiring diagrams depicting the operation of switches were beyond the grasp of automotive technicians. As a result, determining which wires connected to what required deductive reasoning, which turned into a guessing game more often than not.

When answering the following questions, consult Figures 8-5 and 8-6. By reviewing the analysis that follows, your abilities to read and understand wiring diagrams will increase.

1. Suppose the headlights don't operate. What fuse and what color wire powers the headlight switch? What color wire from the headlight switch is used to power the low beam circuit for the headlight? What color wire powers the high beam circuit?

2. If the fuse that powers the backup lights is burned out, what other system(s) would have problems? How does this fuse get power?

3. How does the oil and temp indicator light work? From where does it receive power? If it's a fuse, where is the power source for the fuse? Where is the ground return for the indicator light?

4. How does the stoplight circuit work? What is its power source? How are the stoplights switched on?

Figure 8-7 shows a fuel-injected computerized engine management system for a 1990 Ford E-250 Van. All of Ford's electronic engine control (EEC) sensors and actuators are shown in the diagram. Wiring diagrams for fuel-injected vehicles almost never show what occurs inside the computer electronically, but many do label each wire at the computer's wiring harness. The labels on these wires are often referred to as pin numbers, and each provides hints about what the wires are supposed to do and their expected readings. Some of the following questions assume a basic knowledge of how EFI systems operate—(Chapter 7 on EFI may be used as a reference, if needed.) Answering the following EFI-related questions will help with the transfer of wiring diagram reading skills to actual vehicles. Again, good luck!

1. How do you test the gear select switch for proper operation? What pin numbers at the EEC (computer) do you use for testing? (Hint—Ford sometimes refers to a ground wire as SIG RET, or signal return. Another hint— the squiggly lines inside the gear select switch are resistors.)

2. The EEC power relay has power all the time on the black/orange wire. List the function for each wire going to

this relay. How is the relay controlled? What other components does the relay provide power to?

3. The TPS uses a variable resistor to signal throttle position to the EEC computer. What function does each of the three wires at the TPS perform? What other sensor works the same as the TPS?

4. If the EEC computer cannot supply any of its sensors with a 5-volt reference signal (Pin 26 VREF), then the computer should be replaced. However, before replacing this expensive component, all power inputs and grounds at the EEC computer wiring harness need to be verified. What EEC pin numbers would you have to check for power(s) and ground(s)?

ANSWERS TO WIRING DIAGRAM QUESTIONS
Answers for 1975 Ford Thunderbird in Figures 8-5 and 8-6

Question 1. Power to the headlight switch comes directly from the battery and not through a fuse. A black/orange wire at the starter relay connects a yellow wire that branches out and goes to the fuse panel, ignition switch, and headlight switch. The wire from the headlight switch to the headlights is red/yellow. This wire goes to the beam select switch before it goes to the headlights. The beam select switch directs power to either the high or low beam circuit at each headlight. Power goes to either a red/black wire for the low beams or a light green/black wire for high beams.

Question 2. Finding the fuse that powers the backup lights is the first step in answering this question. The fuses in this diagram are not labeled, so you must start at the backup lights and trace the power wire to the fuse. The two backup lights are on the far right side of the diagram. Each has two wires—one for power, the other for ground. The black wires at each of the backup lights are ground. Notice that many other lights are connected to this wire— thus, the black wire is a common ground. To confirm this, the ground symbol is located on the black wire at the bottom right of the diagram.

The pink/black wire is the power wire for the backup lights and goes to the backup light switch (bottom right in diagram). The wire from the backup light switch to the fuse box is pink/orange and it goes in two directions (indicated by a dot located around the middle left of page two). The left section of pink/orange wire connects to the fuse on the right side (bottom of the box). This fuse receives 12 volts on a gray/yellow wire from the ignition switch. With the key turned ON, the fuse powers up. This same fuse also

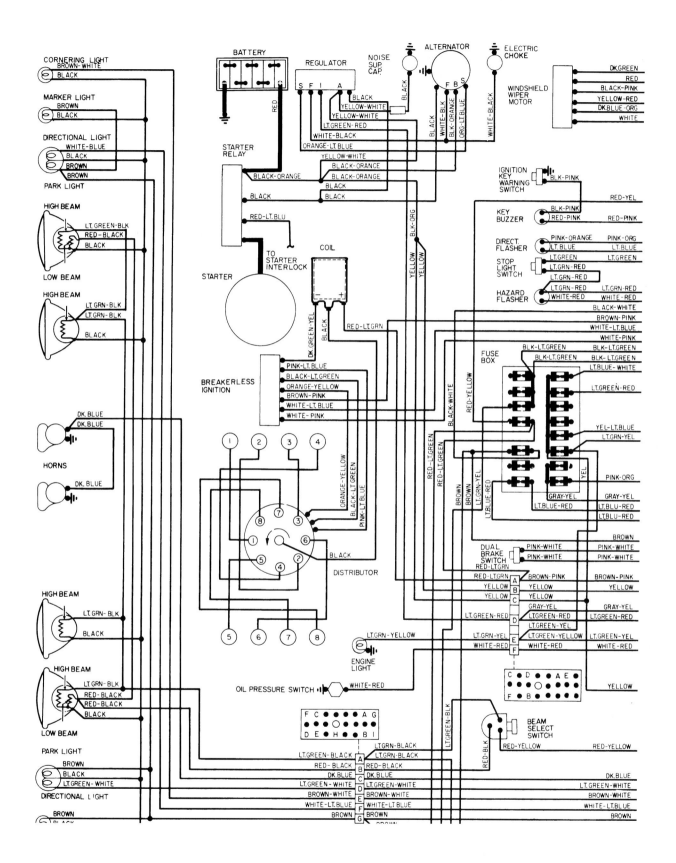

Fig 8-5. *In 1975, the entire wiring diagram for the Ford Thunderbird was contained on only two pages. As vehicles became more complicated, more and more pages were added.* Courtesy of Mitchell 1

YEL-LT.BLUE
T.GRN-YEL

BLACK
LT.GRN-BLK
BLACK
WHITE-RED

LIGHTS
(4 USED)
HIGH BEAM
INDICATOR
DOME

Fig 8-6. *This is the second page of the wiring diagram for the entire circuitry of the 1975 Ford Thunderbird. The first page depicts the front half of the vehicle, including the headlights and engine compartment. Page two exhibits the other half of the vehicle, including the instruments, ignition switch, windshield wiper switch, and rear lighting. Courtesy of Mitchell 1*

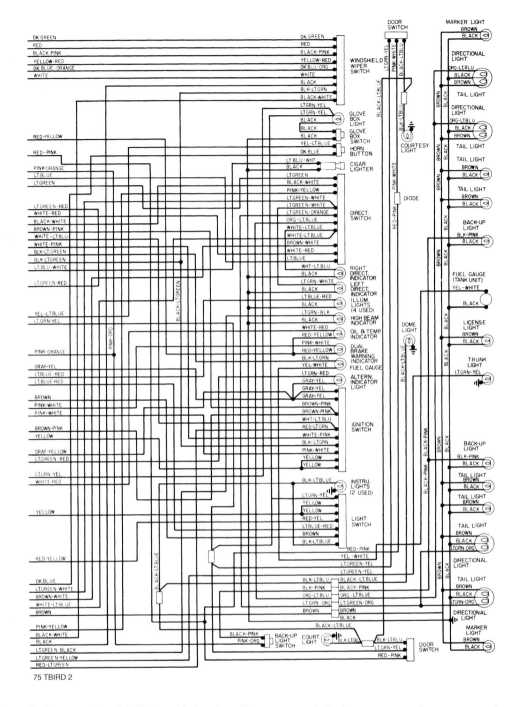

75 TBIRD 2

provides 12 volts to the direct flasher turn signal. Without power, the turn signals will not operate.

Question 3. The oil and temp indicator light is located on page two of the diagram in the center (toward the left). This indicator light is powered by a fuse on the left side of the fuse box (fourth down from the top). The power wire connecting the light to the fuse is red/yellow. This fuse receives 12 volts from the red/light green wire that comes from the ignition switch. The indicator light is controlled

by the oil pressure switch (on page one, bottom center), which grounds the white/red wire (turning on the indicator light) when the ignition is first turned on. After the engine starts, oil pressure opens the switch and breaks the ground connection, and the indicator light goes out.

Question 4. This question is more challenging than previous ones because the diagram doesn't show how the direct switch (turn-signal switch) operates. If the brake lights don't work, you'd need to figure out how the circuit

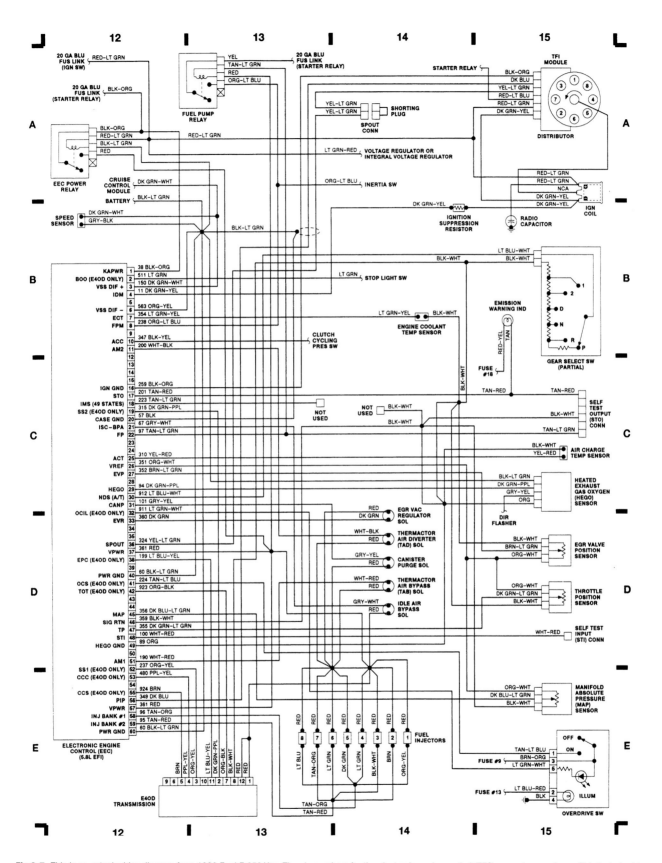

Fig 8-7. *This is an actual wiring diagram for a 1990 Ford E 250 Van. The pin numbers for the electronic engine control (EEC) computer are shown. This typical wiring diagram shows an engine management system.* Courtesy of Mitchell 1

YEL-LT.BLUE
LT.GRN-YEL

BLACK

LT.GRN-BLK
BLACK

WHITE-RED

LIGHTS
(4 USED)

HIGH BEAM
INDICATOR

DOME

BLACK

operates in order to fix it. The fuse that powers the direct switch is located in the fuse box, right side, second fuse from the top. It's connected to the switch via a light green/red wire. When activated, the stoplight switch sends power to the direct switch on the light green wire. Because the diagram doesn't show what happens inside the switch, the assumption is that all four taillights will operate as stoplights when the stoplight switch is closed. Each taillight has a dual-filament bulb (two separate circuits). The right taillight's stoplight wire is orange/light blue and the left is light green/orange; both wires go back to the direct switch. A black wire serves as common ground for all tail- and stoplights. As long as the direct switch has power from the stoplight switch and the wires going from the taillights are connected to the direct switch, the stoplights should work.

Answers for
1990 Ford E-250 Van in Figure 8-7

Question 1. The gear select switch is located at B-15. (Diagram reference numbers run horizontally and letters run vertically.) The light blue/white wire from the gear select switch goes to EEC Pin 30—NDS (A/T), the neutral/drive/safety input for vehicles with automatic transmissions; it prevents engine starting if the transmission is in gear. The other wire from the gear select switch is black/white and goes to several locations including: Pin 8 of the E4OD transmission computer, TPS sensor, EGR valve position sensor, heated oxygen sensor, and EEC Pin 49 (orange wire). Pin 49 is a ground for the heater inside the O_2 sensor. Whenever a wire connects to more than one place, it's probably a ground (though not always!).

To test the switch, connect an ohmmeter to Pins 30 and 49 at the EEC. While watching the ohmmeter, move the gear selector between all gears. Each time gear position is changed, the resistance on the ohmmeter should change. Position 1 should have the lowest resistance and PARK (P) should have highest resistance.

Question 2. The EEC power relay is located at A-12 of the wiring diagram and has a total of four wires. The wires function as follows:
• Black/orange wire—Hot at all times from a fuse link
• Red/light green wire—controls or triggers the relay. It is "hot" from ignition switch in START and RUN positions. It also powers the voltage regulator, ignition coil, and ignition module
• Black/light green wire—provides a ground for the relay's control coil

• Red wire—powers EEC Pins 37 and 67, (VPWR), fuel injectors, five solenoids, and E4OD transmission computer

Question 3. The process of elimination is used to determine the use of each of the three wires at the TPS. The three wires are:
• Orange/white—goes to EEC Pin 26 (VREF), 5 volts or power for the TPS, EGR valve position sensor, and the MAP manifold pressure sensor
• Dark green/light green—goes to TPS signal wire, EEC Pin 47, and TP (the signal that indicates throttle position to the EEC computer)
• Black/white—goes to TPS sensor ground, EEC Pin 48, and SIG RET (signal return). Ford uses SIG RET to represent a sensor ground (hint from question 1)

In this case, the orange/white wire is the power source for the TPS because the dark green/light green wire is the signal wire. Another sensor that works the same as the TPS is the exhaust valve position (EVP).

Question 4. By tracing every wire to the EEC, you'll note there are three power wires and four ground wires. The three power wires are:
• Pin 1 (KAPWR—keep alive power—Hot at all times from a fuse link)
• Pins 37 and 67
• VPWR from the EEC power relay—controlled by ignition switch

The EEC computer also provides grounds to some sensors, but not from the negative battery terminal; they are internal grounds within the computer. These wires don't provide a ground for the EEC computer. The wires that do provide a ground for the EEC are as follows:

• Pin 6 (VSS DIF)
• Pin 20 (CASE GND)
• Pin 40(PWR GND)
• Pin 60 (PWR GND)

All of these wires go directly to the battery's negative terminal. If any of these power or ground wires are not present at the EEC wiring harness, the EEC computer probably won't operate as it should—a possible cause of "no power" on "Pin 26," the 5-volt reference for the computer sensors.

CHAPTER 9
TROUBLESHOOTING ELECTRICAL SYSTEMS

LOGIC? WHAT LOGIC?

Troubleshooting any complex problem requires a logical approach. While this may seem self-evident, many technicians simply don't have a well-thought-out plan of attack when it comes to electrical problems. Oftentimes, they start by disassembling or replacing components while keeping their fingers crossed and hoping for the best. However, if they had only taken the time to carefully analyze an electrical problem and outline the steps required to solve it, they would have found (as you will find after reading this book!) that most electrical problems are really pretty basic.

It's helpful to have a routine procedure when troubleshooting electrical problems. A good habit to develop is to do the easiest and fastest electrical tests first. In general, this should provide the most information with the least amount of effort, and will either solve the problem or at least narrow down its location. However, without a plan, testing wrong circuits and not understanding multimeter readings only leads to frustration and time wasted.

Although you can't see current flowing in a circuit, you can use all your senses to troubleshoot electrical problems, and fortunately, you can also observe the effects of a circuit. For example: lights illuminate, motors spin, relays "click," wires get "hot" when amperage flows, and so forth. Touch, sound, and sight testing methods should all be used simultaneously in conjunction with the use of electrical test equipment.

In addition to using all your senses (and your wits, if you haven't lost them by now), other clues undoubtedly exist about the nature of a particular problem and should be explored. First, determine the conditions under which the electrical problem occurs. Does heat or cold affect the problem? Is it intermittent or does it only happen under specific operating conditions? Did it occur right after an aftermarket accessory was installed or after someone worked on the vehicle? Evaluate the problem(s) to see if any other circuits or components are, or may be, affected. Test other related circuits for proper operation since more than one thing may be wrong. Use the owner's manual and service manual as essential reference materials; these can be invaluable for locating fuses and identifying wiring harnesses, connectors, junction blocks, fusible links, and other electrical components

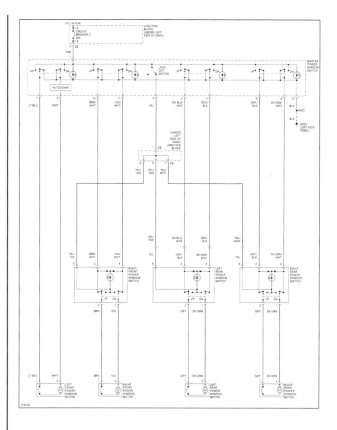

Manufacturers' wiring diagrams are some of the best electrical tools for understanding the nature of any electrical problem. Courtesy of Mitchell 1

located on the same problem circuit. However, the most valuable information about electrical problems can be found on the vehicle's wiring diagram and electrical component locator. These two tools will save you loads of time and provide possible test points within the problem circuit.

"THE STARTER THAT DOESN'T": A CASE STUDY

Consider the following hypothetical scenario: A longtime friend calls to tell you his car won't start. For the past month the starter has cranked slowly, and now it won't turn the engine over at all. Your friend wants to buy a new battery, but you tell him to wait until you can check it out. Now is as good a time as any to put your diagnostic skills to the test, so you head over to inspect his car.

1985 FORD MUSTANG

Component	Component Location
Air conditioning cooling fan control unit	On shake brace, below steering column
Electronic control assembly (ECA)	Under the dashboard
Four-cylinder	On right cowl
V-6 and V-8	Under right front seat
Graphic warning module	At top center of console
Low fuel warning module	Behind right side of dash, above glove box
Speed control amplifier	On left cowl area
Air conditioning blower motor resistor	Under right side of dash, on evaporator assembly
Convertible top circuit breaker	Under instrument panel, to right of steering column
Barometric pressure sensor	On right front fender apron
Coolant temperature sending unit, Four-Cylinder	Cylinder head, front
2.3-liter nonturbo	On rear left side of engine
2.3-liter turbo	On rear left side of engine
V-6	On left side of cylinder head
V-8	Near left side of distributor
Coolant temperature sensor	On top of engine, in front of carburetor
EGR valve position sensor	On top right front of engine
Fuel gauge sending unit	In fuel tank
Low oil level sending unit	On engine oil pan
Air conditioning clutch cycling pressure switch	At top of A/C accumulator
Backup/neutral safety switch	On transmission
Barometric pressure switch	On left shock tower
Clutch safety switch	Above clutch pedal
Dual brake warning switch	On frame rail, near left front shock tower

Some manuals provide electrical component locator charts, like the one shown here. The information in this chart can save you hours of searching for various electrical parts.

Fig 9-1. *When you aren't concerned with how much voltage is present in a circuit, a test light is a quick and easy way to verify that power exists at various test points.*

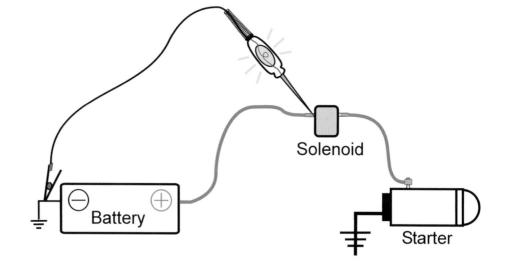

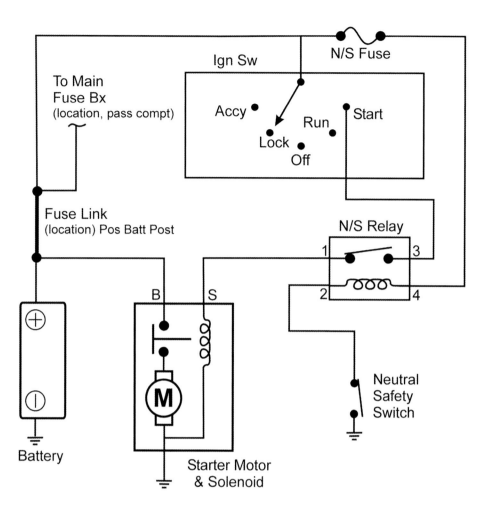

Fig 9-2. *This starter circuit uses a neutral safety switch and relay. These components prevent the starter from cranking the engine unless the transmission is in park or neutral.*

The first step to solving this, or any, electrical problem is to verify the complaint. Then determine if anything else may be wrong. In this hypothetical case, when you try to crank the engine, nothing happens. In addition, no dash lights come on when the ignition key is in the RUN or START position. In fact, nothing seems to be working. Before performing any testing, the battery needs to be checked. Your friend tells you he left the battery on a charger overnight, so it ought to be charged—and he's right. After connecting your DVOM directly to the battery, an open circuit battery voltage is displayed—12.8 volts. This indicates the battery is indeed 100 percent charged—ready to continue testing. Just to be sure, you connect a hand-held battery tester to the battery, and after holding the load switch for 10 seconds, voltage remains above 10 volts, indicating the battery is, in fact, good. So the problem with the car isn't at the battery. Now, it's time to pick up a test light.

Given that you have a charged and tested battery, the next step is to see how far the voltage is traveling within the starter circuit. Using a test light is a quick and easy way

to visually confirm voltage at the starter solenoid. Be sure to always check a test light before testing. (You certainly wouldn't want to spend hours testing wires for power, only to find out later the bulb in a test light is burned out.) With the test light's alligator clip connected to battery ground, you touch test the light's probe to the positive battery terminal. The test light should light up. After connecting the test light as shown in Figure 9-1, the test light should light up when touched to the starter solenoid's battery terminal, confirming voltage at the starter solenoid.

The next step is to find out where the voltage is supposed to be and where it's missing; it's time to take a look at a wiring diagram.

Wiring diagrams show all components of a circuit and how they are connected within the circuit, thereby providing an overview and allowing identification of all possible areas where problems may exist. The wiring diagram in Figure 9-2 shows seven potential trouble spots: 1) battery, 2) starter, 3) fusible link, 4) ignition switch, 5) neutral safety (N/S) switch, 6) N/S relay, and 7) wires connecting all the components together.

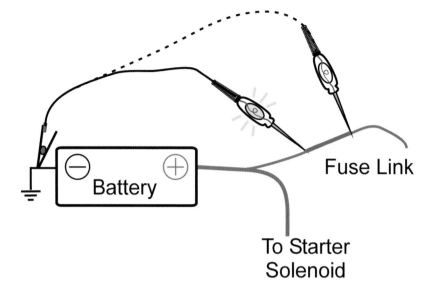

Fig 9-3. *The test light shows that the fusible link is open. Fusible links can disconnect power from a circuit due to amperage overload or just plain old age.*

Fuse Link

To Starter
Solenoid

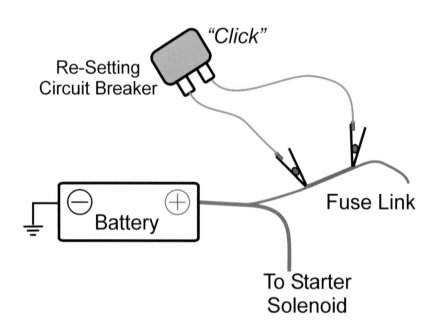

Fig 9-4. *A resetting circuit breaker or a turn-signal flasher unit make good short finders. In our case study, both will cycle off and on as long as they are connected to the fuse link that has shorted to ground.*

"Click"

Re-Setting
Circuit Breaker

Fuse Link

To Starter
Solenoid

The wiring diagram in Figure 9-2 shows a fusible link connected directly to the battery's positive terminal. From here, power goes to the ignition switch. A good rule of thumb for electrical testing is to test the most accessible component(s) first. The ignition switch is under the dash, so some panels have to be removed before access to the wires at the switch can be gained. However, the fusible link is located right next to the battery and it's easy to get to—so, start by testing there. Since we only want to know if voltage is present, continue to use a test light to check for power. Touching the test light probe to the connector on the battery side of the fusible link makes it light up. However, when you touch the probe to the other side of the fuse link, the light doesn't come on, indicating an open fusible link (see Figure 9-3). Problem solved? Maybe.

Fusible links are nothing more than really heavy-duty fuses designed to blow or melt when high amperage conditions exceed their capacity to carry current. In addition to melting, they can wear out from continuous flexing or simply from age. The only way to tell if one has blown because of high current or old age is to replace it. In this case study, after the fuse link is replaced, a large spark is generated when the new fuse link is plugged into the battery terminal harness. Since the ignition key is turned to OFF and

TROUBLESHOOTING ELECTRICAL SYSTEMS

nothing else in the car is turned on, the large spark has to be caused by a direct short-circuit to ground, somewhere on the circuit protected by the fuse link. This short is also the reason why the original fuse link was blown—the cause *not* being old age. Somewhere in the circuit, a power wire—protected by the fuse link—may have worn insulation and its copper strands of wire may have come into contact with ground.

The next step in the discovery process is to find out exactly where the physical problem with the shorted wire is located.

To find out which wire has shorted to ground, the open circuit at the melted fuse link needs to be temporarily reconnected. This can be accomplished via a short finder, as discussed in Chapter 3. A simple short finder can be made with a resettable circuit breaker or turn-signal flasher. The short finder takes the place of the fusible link, cycling on and off as it heats and cools. A turn-signal flasher is actually easier to use for this purpose, since it clicks on and off so you can hear it operate. To narrow the problem, each connector that is part of the circuit powered by the fuse link needs to be unplugged one at a time. If unplugging one stops the short finder from cycling or clicking, then you have found the section of wire that has shorted to ground. As shown in the wiring diagram in Figure 9-2, the fuse link supplies power to the ignition switch and passenger compartment fuse box. When each is unplugged (one at a time!), the short finder keeps cycling or clicking on and off. This means the short still exists. It also means the wires going to the ignition switch or fuse box are not causing the short. This only leaves the wires going from the battery's positive terminal to the fuse link as the location of the short.

After taking a closer look at the wires near the burnt fuse link, some nonfactory wires hiding under the battery are discovered. After asking your friend what they are for, he tells you he had some aftermarket driving lights installed a few months ago. Wires not originally equipped with a vehicle are always suspect as the cause of electrical problems since one can never know how much skill or care was taken during their installation. After pulling the harness out from beneath the battery, you find a red wire (a nonfactory wire); it has been pushed up against the battery tray and there is a section of bare wire where the insulation has worn off from rubbing against the battery tray. After repairing the broken section of wire by replacing it with a new wire and replacing the fuse link again, the ignition switch now turns the instrument lights on and starts the engine. Problem solved? Well, almost.

Unfortunately, the engine still cranks too slowly. If you recall, slow engine-cranking speed was one of the original complaints your friend had all along. There are three common malfunctions that typically can cause a slow-turning starter: (1) excessive engine friction, (2) a physical or electrical problem with the starter, or (3) high resistance in the starter circuit. The first one, excessive engine friction, can be eliminated in this case because once the engine starts, it actually runs smoothly and does not overheat or have smoke coming out the exhaust. (A good method for verifying an absence of excessive friction in the engine is to turn it over by hand using a socket and breaker bar.)

The second potential source of the problem is a little trickier. To eliminate the starter as the problem, you need to know if starter amperage is excessively high or low. If starter amp draw is high, it's most likely caused by a shorted starter field coil or armature; if amperage is low, it's probably caused by unwanted, high resistance somewhere in the starter circuit.

Hopefully by now, you have an inductive ammeter handy to measure starter current draw. First, you disconnect the ignition coil wire going to the distributor so the engine won't start. This will allow enough time for the ammeter to read starter amps when the engine is cranked over. After clamping the current probe around the battery cable at the starter solenoid, the ammeter is ready to measure starter amperage. After cranking the starter, the ammeter reads 82 amps; this seems too low because the engine is a large four-cylinder model and normally requires around 125 amps to start. Furthermore, low amperage in the starter circuit indicates the presence of high resistance. Now it's time to do a voltage drop test with a voltmeter to isolate unwanted, high resistance.

A voltage drop test will determine where the high resistance is located within the starter circuit. The positive side of the circuit needs to be checked first; you connect the red lead of the voltmeter to battery positive and the black lead to the battery cable located at the starter motor. Once the engine is cranked, the resulting voltage drop is 1.4 volts—too high for this type of circuit. In any circuit, the most likely place for high resistance is at a switch.

The switch in this starter circuit is the starter solenoid, which acts like a relay. The ignition switch sends a 12-volt start signal to the solenoid, which in turn connects the positive battery cable directly to the starter. If the contacts inside the solenoid are dirty, they could be the cause of the high resistance, so you connect the voltmeter to both sides of the starter solenoid to repeat the test. Sure enough, the voltage drop is 0.8 volt, indicating high resistance inside the starter solenoid. After replacing the solenoid, the starter cranks at normal speed and the engine starts quickly. Finally, all problems are solved. Case closed.

As can be seen from the foregoing hypothetical case study, a logical approach toward diagnosing electrical

Fig 9-5. *This starter circuit is loosing 1.2 volts somewhere along the positive side of the circuit. By moving the voltmeter's leads along the circuit, the exact location of high resistance can be determined.*

1.2 V

Solenoid

Battery

Starter

This remote starter solenoid can easily be checked for high resistance using a voltage drop test. The contacts inside the solenoid should not exceed a drop of 0.2 volt during starter cranking.

problems is always more productive than a haphazard method. In the sample no-start case, it turned out there were several things wrong with the vehicle. However, the step-by-step approach and methodical checks uncovered all the problems and little time was wasted unnecessarily replacing components. Be sure to always keep in mind the

"Three Things" every circuit needs in order to function—power, load device, and ground return. The previous case study illustrates how to check for three common problems: open circuits, shorts to ground, and unwanted high resistance. In addition to these electrical foul-ups, three other common electrical problems within circuits exist: (1) bad grounds, (2) crossover circuits, and (3) parasitic amperage draw.

BAD GROUNDS

Bad ground connections cause a fair amount of electrical problems. Many technicians have trouble checking for bad grounds because, quite frankly, they don't know exactly what they're looking for. By contrast, checking for power is simple and straightforward—either 12 volts are present or they are not.

There are several ways to check for bad grounds. By far the most accurate method is the use of a voltmeter to perform a voltage drop test as outlined in Chapter 2. Figure 9-6 illustrates how to connect a voltmeter to measure voltage drop on the ground side of a horn circuit. It's important to remember the horn must be operational for a voltage drop test to work. A good ground return will have almost no voltage present. If the horn (the load device in this circuit) hasn't used up all available voltage, the ground connection is bad, and the bad ground is using voltage intended for the horn. Consequently, the horn sounds weak because it doesn't have enough voltage to operate at its full potential.

A test light can be used in place of a voltmeter to check for a bad ground, but it has inherent limitations. When

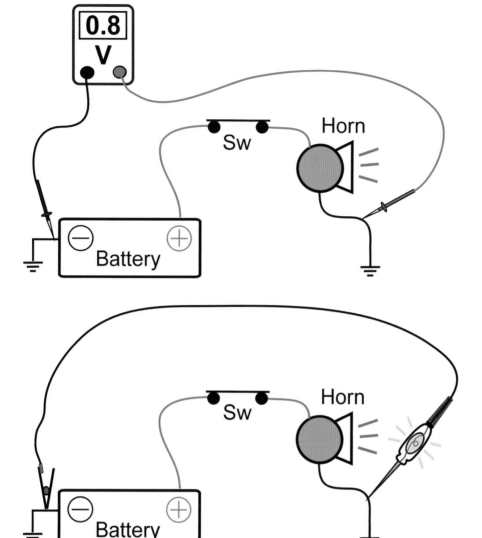

Fig 9-6. *A voltmeter reading of 0.8 volt is too high for the ground return on this horn circuit. The bad ground is stealing voltage from the horn, thus causing it to sound weak.*

Fig 9-7. *A test light can be used to check for a bad ground; however, the ground has to be bad enough for the test light to light up. With this limitation in mind, it's obvious a voltmeter is a better way to find a bad ground.*

touched to the ground side of a circuit, a test light only lights up if there is enough voltage left over to illuminate its bulb. Thus, a bad ground has to have high enough resistance to divide up the voltage between the test light and component being tested in order for a test light to confirm its presence. If the ground is not quite bad enough, the test light won't light up and the presence of a bad ground won't be detected.

Another low-tech method for finding a bad ground is to substitute a ground wire in place of a suspected bad ground. A jumper wire connected between the ground terminal of the load device and a known good ground will help determine if the original ground has high resistance. If the load device works with the jumper wire in place, the ground wire has high resistance. Figure 9-8 (on page 154) illustrates how a jumper wire is used to bypass the ground to the horn.

CROSSOVER CIRCUITS

If a load device is switched on and another unrelated load device also turns on at the same time, a crossover circuit is present. A crossover circuit is created whenever control wires (power or ground side) touch each other. This unwanted contact or connection is usually caused by wire insulation chafing at a connector or inside a wiring harness (where two wires come in contact with each other). Another common cause for crossover circuits is defective electrical component(s). Some circuits are designed so that one component will cause another circuit or component to turn on; consequently, a defect with the controlling electrical part may cause other circuits to operate unintentionally. A final cause of crossover circuits is the idiot electrical technician. When misconnected connectors have been forced together by an unskilled technician, "crossover"

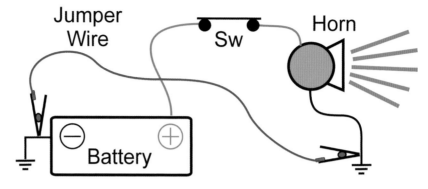

Fig 9-8. *A low-tech, but effective, method for discovering a bad ground is to use a jumper wire as a temporary ground wire. When testing with this method, if the load device works as it should, just repair the ground return wire.*

Jumper Wire

Sw

Horn

Battery

Fig 9-9. *A crossover circuit exists between the backup and trunk lights. The blue/yellow and green wires are touching, or shorted, between the connector and each of the load devices.*

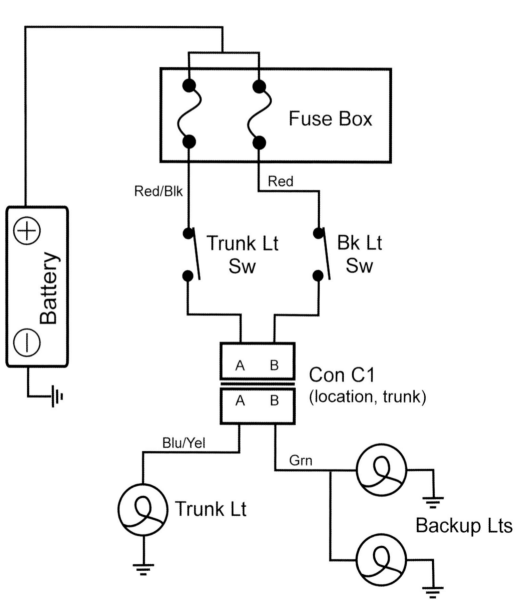

Fuse Box

Red/Blk

Red

Battery

Trunk Lt Sw

Bk Lt Sw

A   B

A   B

Con C1 (location, trunk)

Blu/Yel

Grn

Trunk Lt

Backup Lts

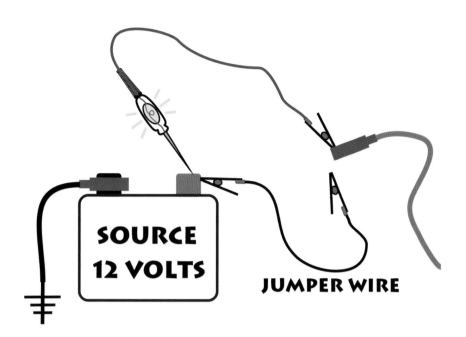

Fig 9-10. *In some vehicles, a computer-controlled relay may be turned on, causing a test light to light up. Using a jumper to temporarily reconnect the battery terminal to the positive battery cable (shown in this drawing) will cause the relay to turn off. Parasitic amp draw testing can be continued once the relay is off.*

SOURCE 12 VOLTS

JUMPER WIRE

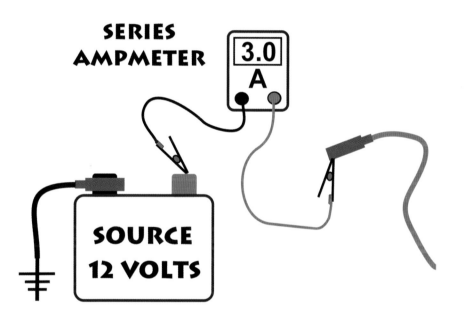

Fig 9-11. *A series ammeter can be connected between either positive or negative battery cables. Make sure the amperage draw is not higher than the meter's capacity to read series amperage, or it could be damaged!*

SERIES AMPMETER

3.0 A

SOURCE 12 VOLTS

circuits occur. Following is an example of an unwanted crossover circuit.

The vehicle in Figure 9-9 has a crossover circuit. Every time the trunk is opened and the trunk light comes on, the backup lights come on as well. The circuit controlling the trunk light is also unintentionally controlling the backup lights. The wiring diagram in Figure 9-9 shows both backup lighting and trunk lighting circuits. As can be seen, the backup lights and trunk light share a common connector labeled C1. By unplugging connector C1 and connecting a continuity tester (a common function on digital multimeters) to

terminals A and B on the power side of the connector (the other half goes to the trunk and backup lights), the existence of a crossover on this side of the connector can be ascertained. If the meter indicates continuity between terminals, the crossover is on that side of the connector. (Always make sure the fuses providing power to the components tested are removed before connecting a tester, since power in the wire where a multimeter is connected can damage the meter during a test for continuity.)

However, if there is no continuity (no beep from the meter) between the two terminals in the connector, then this side of the connector is not the problem. The meter test

TROUBLESHOOTING ELECTRICAL SYSTEMS

leads need to be moved to the other half of the connector. The meter should now beep, confirming the green and blue/yellow wires are touching each other somewhere between the connector and trunk or backup lights. By wiggling the wiring harness while listening to the meter, you can determine exactly where the wires are touching and causing the crossover. The meter will beep intermittently when the wires are wiggled, indicating the location of the bad section of wiring harness. By simply unraveling the wiring harness, you can repair the broken wire insulation with some electrical tape to fix the related electrical problem.

PARASITIC AMP DRAW

Parasitic amp draw occurs often enough that knowing how to diagnose it when it occurs should prove helpful. For example, a car or truck that hasn't been driven for a few days may not start up because of a dead battery. Something may have been left on, drawing power from the battery. This condition is known as a parasitic amperage draw. This problem is commonly caused by trunk, dome, or glove box lights that don't switch off when they should; unfortunately, it usually isn't noticed until it's too late and the battery has been completely drained. Parasitic amp draw can also be caused by a relay stuck on "on." Relays are often used to control power to lighting and EFI computers. A relay that is stuck on "on" while the ignition is in the OFF position will also drain the battery.

Similarly, onboard fuel injection computers, digital radio memory, and other computer-related memory components all use power from the battery to keep their electronic memories alive. Normally, this should not affect a battery's ability to start an engine, even after several months. Standard battery amperage draw should not exceed 75 milliamps (0.075 amps) and is usually less than 40 milliamps. Amperage loads exceeding 75 milliamps will draw current from the battery, leaving it dead after only a few days.

To find a parasitic amp draw, a series ammeter, test light, and jumper wire are needed. Also, some digital multimeters can read series amperage up to 10 amps. These meters are usually (but not always) protected by an internal fuse, so that when connected to a circuit with more than 10 amps, the fuse melts instead of the meter.

How Much Amperage?

Before finding the cause of an unwanted amp draw, you first need to determine approximately how much amperage is being sucked out of the battery. Knowing the amount of amperage draw helps narrow down the type of electrical component that is most likely causing the draw and also helps prevent a blown fuse in a series ammeter.

To do this, first disconnect either the negative or positive battery cable. Connect one end of a test light to the battery terminal and the other to the battery cable. If the amperage draw is high enough (above 4 amps), the test light will light up. On some vehicles, a computer-controlled relay may turn on, causing the test light to light up. To eliminate this occurrence as a potential cause of an amperage draw, temporarily connect a jumper wire between the battery terminal and battery cable while leaving the test light in place (see Figure 9-10). After removing the jumper wire, the test light may go out. If the test light stays on, the amperage draw is above 4 amps. However, be warned! Since you don't know how much in excess of 4 amps are present in the circuit, don't connect a series ammeter to the system as it could cause an internal fuse to blow!

If the test light goes out, the amp draw is less than 4 amps; unfortunately, it may still be high enough to cause a dead battery. To find the parasitic draw, connect the leads of a series ammeter to the battery terminal and battery cable before removing the test light. This will keep the relay (if one is used) from coming on. If the ammeter reads between 1.5 and 0.8 amps, a light of some type is probably stuck on, causing the draw. If amperage is closer to 4 amps, suspect a bad alternator diode as the cause of the draw.

Locating the Parasitic Draw

With either a test light connected in series with the battery terminal and cable (when the draw is above 4 amps), or a series ammeter connected in the same manner (when amp draw is less than 4 amps), the process of locating the source of the problem begins. While watching the test light/ammeter, disconnect the wires going to the alternator. If the light goes out or the meter reading changes, then a leaky diode in the alternator is causing the amperage draw. If disconnecting the alternator doesn't change the test light or ammeter, start removing fuses from the fuse box (under-hood fuse box first, then passenger compartment fuses). If the test light or meter changes when a specific fuse is removed, then the circuit causing the problem has been isolated. Use a wiring diagram to determine which electrical components are powered by the fuse and start unplugging them one at a time—note any changes in the state of the test light or ammeter. If unplugging an ignition key chime or trunk light makes the test light go out or changes the ammeter readings, the parasitic amperage draw has been identified and the problem can be fixed. Oftentimes a relay, switch, or solid-state component needs to be replaced to eliminate the parasitic amperage draw.

SOURCES

I would like to thank the following companies for help with the images and information contained in this book. Without their assistance, many of the images would have been difficult to obtain. All of these companies offer great products and services for both professional and do-it-yourself technicians. Many offer free catalogs or other information on the web; check it out.

FLUKE.

Fluke Corporation

Fluke Corporation is the world leader in the manufacture, distribution and service of electronic test tools and software.

Since its founding in 1948, Fluke has helped define and grow a unique technology market, providing testing and troubleshooting capabilities that have grown to mission critical status in manufacturing and service industries. Every new manufacturing plant, office, hospital, or facility built today represents another potential customer for Fluke products.

From industrial electronic installation, maintenance, and service to precision measurement and quality control, Fluke tools help keep business and industry around the globe up and running. Typical customers and users include automotive technicians, engineers, and computer network professionals. The Fluke brand has a reputation for portability, ruggedness, safety, ease of use and rigid standards of quality.

Fluke Corporation
P.O. Box 9090
6920 Seaway Blvd.
Everett, WA, 98206-9090
800-44-Fluke
www.fluke.com

The First Choice of Automotive Professionals

Mitchell 1

Mitchell 1 began in 1918 with the simple idea that people needed information to repair cars. For more than 85 years Mitchell 1 has provided information solutions to help make automotive professionals' jobs easier. Mitchell Manuals have given way to TeamWorks, the integrated family of software-based solutions anchored by the product that is recognized as the standard for automotive repair information—Mitchell OnDemand.

The TeamWorks family of products is divided into three core components. OnDemand Repair includes the information technicians need to service and repair nearly every car and light truck on the road. OnDemand Repair's factory specifications and procedures, computer diagnostics, electrical wiring diagrams, and detailed illustrations all help technicians fix vehicles fast. OnDemand Estimator gives shop managers and service writers the labor times and OEM parts prices they need to build accurate estimates. OnDemand Manager is a complete business management program that tracks the activities in a shop from estimate to invoice and includes the management tools shop owners need to ensure their businesses run profitably.

TeamWorks' integrated key post-warranty catalogs—including NAPA, Activant, O'Reilly and WORLDPAC—are available to help shops order parts electronically from their preferred vendors saving time and ensures the right parts are delivered the first time.

www.mitchell1.com
888-724-6742

Summit Racing Equipment

Summit Racing Equipment is the leading mail order and Internet supplier of high-performance automotive parts and accessories.

Summit Racing Equipment's catalog is your ticket to the performance industry's lowest prices and biggest selection of parts for street rods, race cars, sport compacts, off-road machines, and more. But the thousands of parts advertised in Summit Racing's catalog are just the beginning of the selection. With access to the entire inventory of hundreds of top-name manufacturers, Summit Racing Equipment can special order the parts you want, even if they're not listed in the catalog!

Ordering your parts doesn't get any easier than at Summit Racing Equipment. You can call the company's toll-free order line at 800-230-3030, 24 hours a day, seven days a week. You can also order from Summit Racing's website, SummitRacing.com. It features an online catalog that is searchable by part number, keyword, vehicle make/model, engine family, manufacturer, and more. You'll also find technical tips, how-to information, charts and guides, and a Q & A section to help customers through their projects. The company offers fast, free shipping on parts delivered via UPS Ground in the continental United States and can provide two-day and next-day delivery service at special rates. With the largest technical department in the industry, Summit Racing Equipment delivers the best tech advice, too.

Summit Racing
800-230-3030
www.summitracing.com

CARQUEST Auto Parts

CARQUEST Auto Parts supplies the professional automotive service repair industry with import, foreign, and domestic auto parts, tools, and equipment. CARQUEST auto parts also are available to motorists who choose to do their own repairs, though due to the advanced technology in today's automotive vehicles, it is CARQUEST's recommendation that auto repairs be done by professional automotive service technicians. For the location of the CARQUEST Auto Parts Stores nearest you, visit: www.CARQUEST.com or call 800-492-PART.

Weller

Weller, A Division of Cooper Hand Tools

Weller has come a long way from its humble beginning back in 1945 when Carl Weller, a home radio repairman, wasn't satisfied with the bulky, slow-heating soldering irons that were the tools of his trade. Today, the Weller brand is still synonymous with electronics design innovation. Weller is now the most respected name in specialized soldering and desoldering products for production, rework, and repair of through-hole and SMT boards, products for contact removal of ICs and QFPs, noncontact hot air products, and much more. Weller's latest line of products again breaks new ground; this time in the fast-growing arena of lead-free soldering technology. Weller Silver Series stations have the power to handle the extra demands of lead-free soldering and the versatility to tackle a wide array of applications.

3535 Glenwood Avenue
Raleigh, NC 27612
(919) 781-7200
www.cooperhandtools.com

SPX/OTC

SPX/OTC is a major manufacture and supplier of vehicle electronic diagnostic instruments, automotive fuel system maintenance equipment, special service tools, general purpose tools, pullers, heavy duty tools, shop equipment, and hydraulic components. OTC's full range of tools and equipment has been designed for the professional service technician and represents the highest standard in quality and reliability. OTC-brand tools are engineered in the United States and backed by a Lifetime Marathon Warranty.

SPX Service Solutions provides specialty service tools for motor vehicle manufacturers' dealership networks, integrated service, technical, and training information for vehicle original equipment manufacturers and professional test and service tools for the HVAC, refrigeration, and electrical trades.

655 Eisenhower Drive
Owatonna, MN 55060
800-533-5338
www.otctools.com

Northwest Regulator

Northwest Regulator Supply, Inc. has been in the automotive electrical business since 1961. It's primary focus is the development, manufacturing, and marketing of automotive equipment (under the AmFor Electronics name); direct connection test leads; and repair, adapter, and custom wiring harnesses for automotive and other types of electronic applications.

800-242-6367
www.nwreg.com

INDEX